Farm Policy and Trade Conflict

Thames Essays on Contemporary International Economic Issues

Published under the aegis of the Trade Policy Research Centre, London

Books in this series focus on key topics affecting the international economy, including trade, the activities of multinational enterprises, and international macroeconomics. They are written by leading academics and practitioners in international economics and addressed to the business and policy-making communities as well as the academic market. Each essay covers an issue of topical concern.

Titles in the Series:

The Technological Competitiveness of Japanese Multinationals,
  by Robert Pearce and Marina Papanastassiou

Farm Policy and Trade Conflict: The Uruguay Round and CAP Reform,
  by Alan Swinbank and Carolyn Tanner

Farm Policy and Trade Conflict

The Uruguay Round and CAP Reform

Alan Swinbank and Carolyn Tanner

Ann Arbor

THE UNIVERSITY OF MICHIGAN PRESS

Copyright © by the University of Michigan 1996
All rights reserved
Published in the United States of America by
The University of Michigan Press
Manufactured in the United States of America
⊗ Printed on acid-free paper

1999   1998   1997   1996   4   3   2   1

A CIP catalog record for this book is available from the British Library

Library of Congress Cataloging-in-Publication Data

Swinbank, A. (Alan)
    Farm policy and trade conflict : the Uruguay Round and CAP reform
/ Alan Swinbank and Carolyn Tanner.
        p.   cm. — (Thames essays)
    Includes bibliographical references and index.
    ISBN 0-472-10727-5 (acid-free paper)
    1. Agriculture and state—European Union countries.   2. Produce
trade—Government policy—European Union countries.   3. Uruguay
Round (1987–1994)   I. Tanner, Carolyn.   II. Title.   III. Series.
HD1918.S95   1996
338.1'84—dc20   96-10155
                                                               CIP

Contents

Preface

This Thames Essay is concerned with one facet of the Uruguay Round of the General Agreement on Tariffs and Trade (GATT) negotiations that were launched at Punta del Este in September 1986 and, after several extensions, were concluded in December 1993. In particular, it focuses on one of the GATT dossiers, that relating to agriculture, and briefly considers sanitary and phytosanitary measures. Moreover, it is primarily concerned with the European Community's (EC) common agricultural policy (CAP) and U.S.–EC relations. Agriculture, of course, was not the only contentious dossier and nor were the EC and the United States the only actors. During the negotiations, important coalitions and conflicts arose between the GATT members, and formal and informal linkages of policy issues were woven.

The GATT negotiations were often overshadowed by other domestic and international policy concerns. These too have played their part in the evolving drama, in some instances diverting attention away from the GATT discussions toward more urgent matters of policy concern and in other instances weakening or strengthening alliances.

In 1986, President Ronald Reagan of the United States was halfway through his second term of office, and the Soviet Union still ruled its empire of subservient states. By December 1993, Reagan's presidential baton had been handed on—first to George Bush and then to Bill Clinton—while Mikhail Gorbachev had been overthrown. His successor, Boris Yeltsin, had sacked the Russian Parliament, the Soviet Union had fragmented, and the former satellite states of central and eastern Europe had gained independence, and several were clamoring for membership in the EC. In the former Yugoslavia, bitter civil wars within and between the states had impoverished much of the region.

These dramatic events dominated European politics at the end of the 1980s and inevitably relegated the GATT talks to a lower priority. Perestroika, and the general relaxing of tension along the Iron Curtain, meant that Austrian, Swedish, and Finnish membership in the EC had become politically possible, and, indeed, in 1993 accession negotiations began with these three states and another hopeful, Norway. The tensions these developments occasioned within the EC, between those who saw "deepening" as an alternative to "widening" of the EC, were considerable. The political independence of the Democratic Republic of Germany (East Germany) and its subsequent absorption into the Federal Republic of Germany (West Germany) in October 1990 expanded the EC at a stroke, but not without costs. German preoccupation with the social and economic costs of the integration of its eastern and western parts was understandable. However, the Bundesbank pursued a high interest rate policy in its attempt to constrain the inflationary impact of the substantial budget deficits incurred by the Federal Republic in funding the revitalization of the eastern *Länder*. This policy placed considerable strains upon the exchange rate mechanism (ERM) of the European Monetary System (EMS). The ERM had been seen as a forerunner of the Economic and Monetary Union (EMU) in the Maastricht Treaty on European Union, but the system collapsed in August 1993. In the former Yugoslavia, the Bosnian conflict had exposed as bogus the EC's aspirations to forge a common foreign policy.

In August 1990, Iraq invaded Kuwait and in February 1991 was expelled from the country by a United Nations (UN) task force headed by the United States. In December 1990, it had been feared that the collapse of the GATT talks in Brussels might lead to a trade war, with the EC and the United States as the principal adversaries. The outbreak of the Gulf War accentuated the common interests of the United States and its EC partners, and in the aftermath of the conflict in the Gulf quiet diplomacy resuscitated the GATT talks.

On 1 January 1986 the EC was enlarged to embrace Portugal and Spain. Almost immediately, as the CAP began to be applied in Spain, an import levy was placed on maize and other cereal imports into the country. This, in American eyes, was a violation of their trading rights: in their view, EC enlargement should have led to Article XXIV:6 negotiations in the GATT talks, offering compensation to trading partners who were likely to suffer loss as a result of enlargement. This U.S.–EC trade dispute, together with simmering U.S. discontent with the EC's preferential access arrangements for citrus and its complaints over the EC's export subsidies on pasta, inevitably resulted in the United States and the EC distrusting each other at the outset of the negotiations.

But the trading implications of enlargement were not the only EC trade policy arena in which the United States distrusted the EC's motives. From its very inception it had always been intended that the EC should become a single "common" market. However, breaking down all barriers to intra-Community trade had proved difficult. In 1985, the Commission of the European Communities launched a new political initiative to remove the remaining barriers to intra-Community trade; and in the Single European Act of 1987 this took shape as the EC's 1992 Programme. While American concerns that this amounted to a "Fortress Europe" were undoubtedly exaggerated, the EC's preoccupations in the late 1980s were certainly focused on internal trade liberalization rather than external trade liberalization. Worries about the trade implications of the 1992 Programme, combined with the weakening of Soviet control over central Europe, prompted the European Free Trade Area (EFTA) countries to seek special trading links with the EC in a European Economic Area (EEA) and caused several of them to apply for membership in the EC. Nonetheless, the enlargement negotiations with the alpine and nordic applicants were held up until the 1992 Programme could be said to be complete.

The apparent success of the Single European Act and the EC's 1992 Programme encouraged the integrationist camp within the EC, and particularly the then president of the Commission of the European Communities, Jacques Delors, to press for further constitutional change. Chancellor Helmut Kohl of Germany and President François Mitterand of France were the other key proponents. Thus, intergovernmental conferences were launched that culminated in the initialing of a Treaty on European Union in Maastricht in December 1991. While some of the debates within the intergovernmental conferences had been tense, the subsequent battles within the member states as their governments sought ratification of the Maastricht Treaty threatened the political stability of the EC. Debate between pro- and anti-Maastricht lobbies was particularly fierce in Denmark, France, and the United Kingdom and to a lesser extent in Germany. Not only did these preoccupations with constitutional issues crowd out other topics—such as the GATT talks—from the EC's agenda, but, with concerns about national sovereignty now at the fore, there was a positive reluctance on the part of the majority to impose a GATT solution on a reluctant minority. Thus, when the European Council met in Brussels on 29 October 1993 to "celebrate" the coming into force of the Maastricht Treaty on 1 November 1993, GATT was not even on the agenda—although everyone knew that a "final" deadline of 15 December 1993 had been set for the conclusion of the talks and that France was still implacably opposed to the proposed outcome on agriculture.

As 1993 progressed, the United States was also challenged by other foreign policy concerns that tended to eclipse the GATT talks. Although a U.S.–Canada free trade area agreement had already been put in place, the proposal to extend this to a North American Free Trade Agreement (NAFTA), embracing Mexico as well as the United States and Canada, aroused vocal opposition. In the run-up to the House of Representatives' vote in November 1993 the Clinton administration had very little room for maneuver on the technical issues that still had to be resolved between the United States and the EC if the Uruguay Round was to be brought to a successful conclusion. Furthermore, a rejection of NAFTA—which at one time seemed to be a distinct possibility—would have strengthened the position of the American opponents to the emerging Uruguay Round agreement and placed its passage through Congress at risk. On 17 November 1993, NAFTA was passed by the House of Representatives by a surprisingly large margin of 234 votes for and 200 against; and with just four weeks to run before the final GATT deadline of 15 December 1993, the closing phase of the Uruguay Round negotiations could begin.

The European Union

On 25 March 1957 in Rome, six West European nations (Belgium, West Germany, France, Italy, Luxembourg, and the Netherlands) signed the Treaty establishing a European Economic Community (EEC), which came into force on 1 January 1958. It was this treaty that provided for the establishment of the CAP described in these pages as well as for a common commercial policy that empowered EEC institutions to act on behalf of the member states in GATT negotiations.

There were two other European communities to which these six European states belonged: the European Coal and Steel Community (ECSC) and the European Atomic Energy Community (EURATOM). They shared common institutions: namely, the two bodies that we will encounter most frequently in this text—the Council of the European Communities and the Commission of the European Communities. Somewhat inaccurately—though understandably—the European Communities were frequently referred to in the singular as the European Community (EC).

These treaties have been amended on a number of occasions: for example, by the Merger Treaty of 1967 and by the Treaties of Accession for Denmark, Ireland, and the United Kingdom in 1973; for Greece in 1981; and for Portugal and Spain in 1986. More substantial changes came with the Single European Act of 1987 and the Maastricht Treaty on European Union agreed upon in December 1991, which entered into force on 1 November 1993.

The Maastricht Treaty did not replace the ECSC, EEC, or EURATOM treaties, but it did strike the word *Economic* from the title of the EEC Treaty, which is now known as the "Treaty establishing a European Community." However, Article 38 of this treaty still provides for the CAP, and Article 113 empowers the Commission to negotiate on behalf of the European Union (EU) in GATT. In addition to amending the three Treaties establishing the European Communities, the Maastricht Treaty added two further "pillars" to the Union: that of a common foreign and security policy and that stipulating cooperation in the fields of justice and home affairs. On 1 November 1993, the name European Union was assumed, and this "new" entity includes the three European Communities and in particular the rechristened European Community (EC) on which the CAP and the GATT negotiations are based. The Council of the European Communities has assumed the title "Council of the European Union," and the Commission of the European Communities, the title "European Commission."

Acknowledgments

This text had its genesis in a period of sabbatical leave that Carolyn Tanner spent in the Department of Agricultural Economics and Management at The University of Reading in 1992 and a paper she subsequently presented to the thirty-seventh annual conference of the Australian Agricultural Economics Society held at The University of Sydney in February 1993 (Swinbank and Tanner 1993). The project was substantially furthered when Alan Swinbank was visiting professor in the Department of Agricultural Economics at The University of Sydney in August 1994. In truth, however, it has evolved over a good many years, reflecting discussions with colleagues and students in our respective institutions and elsewhere as well as the ideas advanced in conference and journal papers by our friends around the world. We are particularly grateful to Simon Harris for his help and to the Right Honourable Paul Channon MP for his time. As will be evident as you read through the text, the weekly journal *Agra Europe* has been a particularly useful source.

Chapter 4 incorporates much of the material presented at a conference at the University of Warwick in March 1992 (Swinbank 1992), and we have borrowed heavily from other presentations and papers of recent years (in particular, Tanner and Swinbank 1987; Swinbank 1993a; and Swinbank 1994). Much of chapter 3 is adapted from a module manual written for The University of Reading MBA program by Alan Swinbank in 1993.

Finally, we should express our thanks to Allan Webster and

Matthew McQueen of the Trade Policy Research Centre (TPRC) for inviting us to write this Thames Essay. It is, of course, the case that the views expressed in this monograph, and any errors or omissions, are those of the authors and not those of TPRC, The University of Reading, or The University of Sydney.

CHAPTER 1

Agriculture in the GATT

The Uruguay Round was the eighth in a series of multilateral trade negotiations held under the auspices of the General Agreement on Tariffs and Trade (GATT). The round was launched by a ministerial declaration in Punta del Este in Uruguay on 21 September 1986. This was clearly the most ambitious of the GATT declarations to date. Compared with earlier rounds, the Punta del Este Declaration was much broader in its aims because, for the first time, trade in services and investment was included and, more importantly for the purposes of this text, agricultural products received special mention.

During much of the 1980s, agricultural trade had been affected by the accumulation of surplus stocks (especially in the United States and the EC). In an effort to clear stocks, governments resorted to export subsidies, which depressed agricultural prices further. This was a critical time for agricultural trade, and we will discuss the prevailing conditions of this period in chapter 2. In addition to problems in the agricultural sector, the multilateral trading system as a whole was in crisis. The increasing use of nontariff barriers and antidumping measures became a source of growing tension between the major trading partners and, as a consequence, there was an erosion of respect for the GATT rules. The lack of confidence in the multilateral trading system led to a proliferation of preferential trading arrangements during the 1980s, further undermining the multilateral system.

The purpose of this chapter is to provide a brief review of the development of the GATT, the separate treatment of agriculture under the GATT rules, and the main outcomes of the earlier negotiating rounds as they affected agriculture.

The General Agreement on Tariffs and Trade

The establishment of the GATT was driven by the concern to avoid a repetition of the disastrous mistakes of the 1920s and 1930s. Midway through the Second World War, the United States and the United Kingdom discussed plans for a postwar framework of international economic cooperation involving three organizations that would operate as agencies of the UN. Two of these organizations—the International Monetary Fund and the International Bank for Reconstruction and Development (or the World Bank as it has become known)—came into being. However, the third—the International Trade Organization (ITO)—barely saw the light of day. Despite the failure of the ITO, it was to prove historically significant in that preparations for its establishment resulted in the creation of the GATT.

Historical Background

As a result of the bilateral talks between U.S. and U.K. officials to establish a multilateral, nondiscriminatory trading system, the U.S. Department of State in 1945 published a set of proposals for an international trade organization. The proposals were subsequently elaborated into a draft charter that formed the basis for successive UN-sponsored conferences held from 1946 to 1948 in London, Lake Success (New York), Geneva, and Havana. The final version of the ITO Charter, drawn up in Havana, became known as the Havana Charter.

At the first meeting of the UN's Economic and Social Committee in February 1946, a U.S. initiative calling for an international conference to consider the establishment of the ITO was endorsed, and a Preparatory Committee to prepare for the conference and draft the ITO Charter was set up. When the committee met in London later that year, the idea of holding multilateral talks aimed at reducing tariffs, prior to setting up the ITO, was canvassed among member countries. Prior to the conclusion of the conference, the United States formally invited all countries interested in negotiating with the United States to gather in Geneva in April 1947.[1]

To provide a framework for the tariff negotiations, the Preparatory Committee outlined a set of procedures to be followed. At the next meeting at Lake Success, from January to February 1947, the procedures for the tariff negotiations were developed and a draft of the GATT was drawn up for signing at the Geneva conference commencing in April of that year. To ensure that the tariff concessions that were negotiated were not undermined by other trade measures, the Geneva agreement incorporated many of the commercial trade provisions of the draft ITO Charter.

---

1. For discussion of the background to the U.S. offer to negotiate tariff reductions, see Curzon and Curzon (1976, 143–46).

When the Preparatory Committee met again in Geneva in April 1947, it completed the preparations for establishing the ITO and, more significantly as it later transpired, held the first round of multilateral trade negotiations. On 30 October 1947 at the end of the conference, the twenty-three participants signed the GATT, which contained a provisional agreement on trade relations among the signatory countries. By the terms of a Protocol of Provisional Application signed at the same time by eight participating countries, the General Agreement was to be applied beginning on 1 January 1948, pending the establishment of the ITO. The GATT was envisaged as an interim measure only, with the secretariat to be provided by the ITO.

The UN's Conference on Trade and Employment commenced in Havana in November 1947 with fifty-six countries participating. The extent of the conflict between the representatives can be gauged by the length of the conference (which ran until March 1948) as well as by the number of amendments (602 in all) that were added (*Trade Policies for a Better Future* 1987, 160).[2] The final compromise agreement, which was signed by fifty-three countries, provided for the establishment of the ITO. Controversy continued to surround the ITO, particularly in the United States. By mid-1950, only two countries—Australia and Liberia—had ratified the Charter. When the Truman administration announced in December 1948 that it would not submit the ITO Charter to Congress for ratification, efforts to establish the ITO were abandoned.

Meanwhile, the GATT continued to operate—the second round of multilateral trade negotiations was held at Annecy (France) in 1949—and it assumed the commercial policy role that had previously been assigned to the ITO. As Gerard Curzon (1965, 32) observed: "By the fortuity of historical and political circumstances, the essence of multilateralism survived without any of the rigid rules that might have limited its efficiency." A multilateral approach to trade liberalization is to be preferred to other approaches on both economic and political grounds. Economists would argue that, under most circumstances, unilateral trade liberalization will raise a country's economic welfare by reducing economic inefficiency and stimulating competition (Baldwin 1987, 37). Nevertheless, politicians wishing to implement unilateral liberalization have to contend with two things: the political influence of the protected industry, which is usually better organized and far outweighs that of the consumer group and the widely held mercantilist belief that exporting is

---

2. For discussion of some of the major areas of conflict, see Curzon (1965, chapter 1), Curzon and Curzon (1976, 143–46), and Dam (1970, chapter 2).

good for a country and importing is bad (see Baldwin 1987, 38). It can be shown that a multilateral approach yields benefits beyond those of unilateral liberalization due to the expansion of world trade and the opportunity for increased specialization. Multilateral liberalization has the added political advantages of mobilizing exporters to support liberalization (thus helping to offset the protectionist lobbying) and of helping to promote more harmonious international trade relations.[3]

But the GATT was not ratified by the signatory countries or contracting parties, as they are termed. Countries waited to see if the agreement would be ratified by the United States. In the view of Gerard and Victoria Curzon (1976, 146), "the inner tensions prevailing on the domestic political front in the United States" were such that had "GATT been submitted to the Senate for ratification, it would almost certainly have failed to receive the necessary two-thirds majority. As a result, the whole of GATT is applied *provisionally,* not only in the United States but by all other contracting parties as well." Despite this inauspicious start, the membership of the GATT has continued to grow, numbering over eighty countries by the early 1970s and over one hundred by the early 1990s.

Principles of the General Agreement

The General Agreement contained a framework for the mutual reduction in tariffs between the contracting parties and a code of conduct regulating governmental interference in international trade. The preamble to the General Agreement set out the economic objectives as follows:

> raising standards of living, ensuring full employment and a large and steadily growing volume of real income and effective demand, developing the full use of the resources of the world and expanding the production and exchange of goods. (Dam 1970, 391)

According to the preamble, these objectives were to be achieved by the contracting parties "entering into reciprocal and mutually advantageous arrangements directed to the substantial reduction of tariffs and other barriers to trade and to the elimination of discriminatory treatment in international commerce" (Dam 1970, 391). As noted by McGovern

---

3. Other approaches to trade liberalization and negotiations are discussed in Baldwin (1987).

(1986, 12), it is unlikely that the signatories intended the "fairly cursory statement of ends and means" provided in the preamble to be "comprehensive," and other principles can be inferred from the terms of the General Agreement.[4] One legal expert, Frieder Roessler (1987, 71–72), while acknowledging the complexity of the codes of conduct, has identified three basic underlying principles:

- *Nondiscrimination* or the most-favored-nation (MFN) principle requires each member country to treat trade of all other member countries equally; that is, any advantage given to one member must be given immediately and unconditionally to all other GATT members;
- *Open markets* or free trade: this "principle is realised by the General Agreement through a prohibition of all forms of protection except customs tariffs and the establishment of a procedural framework for tariff negotiations"; and
- *Fair trade:* this principle is encompassed in the General Agreement's prohibition of the use of export subsidies on manufactured products and limitation of their use for primary products. "While the General Agreement does not prohibit dumping and domestic subsidies, all member countries have the right to levy anti-dumping or subsidy-countervailing duties if their industries are materially injured by goods dumped in their market or entering with the help of subsidies." (Roessler, 72)

Another commonly invoked principle is that of *reciprocity.* For example, in the view of the Curzons (1976, 156), reciprocity is "at the core of the international trade system as it has evolved since 1947." Although the articles of the General Agreement neither define reciprocity nor how reciprocal negotiations are to be conducted, in practice, reciprocity has become a fundamental element in the modus operandi of the GATT.[5] As Winters has noted (1987, 45),

Reciprocity in trade negotiations comes in many guises, ranging from the simple bilateral swapping of tariff reductions, through

---

    4. For a discussion of the legal constitution of GATT and the development of the rules, see Dam (1970), Hudec (1975), and McGovern (1986).

    5. A contrary view is held by McGovern (1986, 13), who argues that reciprocity "seems little more than an aspect of self-interest" in that countries "will naturally prefer to give concessions as part of a bargain in which they acquire benefits in return."

multilateral reciprocity, to qualitative reciprocity in which mutual concessions are made in the design of agenda or codes of practice. Although reciprocity is difficult to define, several conventions grew up during the early GATT rounds to make it operational—such conventions as the use of trade coverage to measure it and the adoption of principal–supplier modes of negotiations to arrange it. . . . Qualitative reciprocity is already evident in the agenda for the Uruguay Round.

Practical politics has demonstrated that trade liberalization can best be achieved in a multilateral (or bilateral) framework in which reciprocal benefits are traded. Reciprocity also implies that trade-offs can be brokered between different trade sectors, but the EC has tended to argue that agriculture should be treated in isolation, thus avoiding the need to make concessions on agriculture to obtain reciprocal access to protected markets for services or manufactured goods. Over the years it has been agreed that developing countries should receive "special and differential treatment" and thus be excused from reciprocity.

The general principles of the GATT are meant to apply, with only limited exceptions, to *all* trade. The most important exception to the MFN principle is the one set out in Article XXIV, which permits countries to form customs unions or free trade areas provided that the trade barriers following integration are not on the whole "higher or more restrictive than the general incidence of the duties and regulations of commerce" that were applied by the constituent countries prior to integration. However, where an increase in duty is proposed for a bound item, the GATT rules provide for compensation, but "due account must be taken of the compensation already afforded by the reductions brought about in the corresponding duties of other constituents of the union" (McGovern 1986, 265). These provisions of the GATT formed the basis of the "chicken war" between the United States and the EC in the 1960s, which we will discuss later in this chapter. Another important exception to the MFN principle, which has applied since 1971, is the Generalized System of Preferences whereby developed countries grant tariff preferences to developing countries (Roessler 1987, 72).

There are also exceptions to the principle of open markets or free trade. For example, if the tariff commitments or any other obligations under GATT have led, or threaten to lead, to serious injury to domestic producers, then a member country is permitted to take emergency action on imports of that product. The General Agreement also permits member countries to impose quantitative restrictions to deal with balance-of-payments difficulties, provided such restrictions are gradually relaxed as the balance-of-payments situation improves.

The Application of the GATT Rules to Agriculture

It is generally recognized that the GATT rules relating to trade treat agriculture differently from other industries and that, from the outset, agriculture has been a problem area for the GATT system for liberalizing world trade.[6] A major reason for the failure of the United States to ratify the ITO Charter was that it was incompatible with U.S. agricultural policy at the time. As one of the major participants in the negotiations leading to the GATT, the United States was in a strong position to ensure that U.S. agricultural programs would not be threatened. From the outset, the approach to agricultural trade and the approach to manufactured trade in the GATT have been fundamentally different. As Hathaway (1987, 103–4) has noted:

> In general, GATT rules relate to how governments may intervene to protect domestic markets and industries. . . . These rules were agreed to by member countries of the GATT, and governments brought their practices in line with these rules.
>
> For agriculture, the process was exactly the reverse. The GATT rules were written to fit the agricultural programs then in existence, especially in the United States. Since then the rules have been adopted or interpreted to fit various other national agricultural policies. So instead of developing domestic agricultural policies to fit the rules of international trade, we have tried to develop rules to fit the policies.

It is not surprising, therefore, that the GATT rules as they have evolved have failed to provide a suitable framework for the conduct of agricultural trade or for negotiations on agricultural trade issues. There are two main areas in which agricultural trade receives special treatment: subsidies (Article XVI) and quantitative restrictions (Articles XI and XIII). These areas of special treatment have also been the basis for many trade disputes between contracting parties to the GATT.[7]

The original GATT rules did not prohibit the use of subsidies in

---

6. See, for example, Hathaway (1987) and Johnson (1991, chapter 13).

7. For a discussion of agricultural trade disputes, see Curzon and Curzon (1976, chapter 3), and Hathaway (1987, chapter 5).

general. The contracting parties were merely required to notify other countries of any subsidies that directly or indirectly reduced imports or increased exports. In recognition of the undesirable effects of export subsidies, Article XVI was extended, in 1955, to include what Johnson (1991, 313) has referred to as the "totally meaningless provision that export subsidies were not to be used to achieve 'more than an equitable share of world trade in that product.'" Needless to say, there has been much debate on what constitutes an "equitable share," but more fundamentally the provision implies a market share approach to world trade that is not acceptable to all countries. Additional influence of U.S. domestic agricultural policy on the framing of the GATT rules can be seen in the U.S. refusal, in 1958, to endorse a total prohibition on the use of subsidies.

As Hathaway (1987, 106) has noted, when the provision that specifically prohibits the use of export subsidies on all products other than primary products was made part of Article XVI, "the separate treatment of agriculture was complete." During the Tokyo Round of negotiations, completed in 1979, an attempt was made to devise a subsidies code that could be enforced. However, enforcement proved elusive.

The other major area in which the treatment of agriculture is different from that of other products is in quantitative restrictions. As with subsidies, the treatment of agriculture under the quantitative restriction rules was developed to accommodate U.S. domestic agricultural policy and has been controversial from the outset. The agricultural exceptions to Article XI, which prohibits the use of quotas, are as follows:

- Export restrictions can be used to prevent or relieve critical shortages of foodstuffs or other products essential to the exporting country.
- Import and export restrictions can be used to bring about "the application of standards or regulations for the classification, grading, or marketing of commodities in international trade."
- Import restrictions may be applied on any agricultural or fisheries product imported in any form necessary to the enforcement of governmental measures that operate to: restrict the production or marketing of the like domestic product or of a domestic product that is a close substitute; remove a temporary surplus of a like domestic product by making the surplus available to groups of domestic consumers free or at reduced prices; or restrict the quantities produced of any animal product that is directly dependent wholly or mainly on the imported product. (Hathaway 1987, 108–9)

Despite these exceptions, the United States had difficulty accommo-dating the GATT rules, and in 1951 Congress amended Section 22 of the Agricultural Adjustment Act of 1933 to include the following statement: "No trade agreement or other international agreement heretofore or hereafter entered into by the United States shall be applied in a manner inconsistent with this section" (Johnson 1991, 311–12). Section 22 re-quired the administration to impose quantitative restrictions (or special fees) in cases where imports would impinge upon the effectiveness of a farm program. Following the passage of this amendment, import quotas were imposed on wheat and other grains, cotton, peanuts, and dairy products.

For all products other than dairy, U.S. farm programs were in place that restricted production, and the effect on imports was probably consis-tent with Article XI. In the case of dairy products, the problem was that domestic production was not restricted, and import quotas were set at very low levels. In 1951 and in each of the three subsequent years, the U.S. dairy import restrictions were found to infringe Article XI, and the United States was subject to retaliatory provisions. When the GATT underwent an overall review in 1955, the United States sought to legalize its position with respect to Section 22 quotas and requested a formal waiver.[8] Threats by the United States that it would otherwise be forced to withdraw from the GATT and concern by other countries that, as their balance-of-payments situation improved, they would no longer be able to impose quotas resulted in the granting of a "temporary" waiver to the United States.[9] The granting of the waiver released the United States from its obligations under Article II with respect to fees and Article XI with respect to quotas for actions taken under Section 22 (Warley 1976, chap. 3). The waiver applied not only to dairy products but to all agricultural products, irrespective of whether they were sub-ject to supply controls under domestic programs. Unlike the later waiv-ers granted to Germany and Belgium, the U.S. waiver was without time limit and was subject only to a requirement to report annually to the GATT. The waiver meant that the United States was free to introduce import quotas on *any* agricultural product.

The effect of granting the waiver on the trading system was summed up by Dam (1970, 260–61):

---

8. For discussion of the rules granting a waiver from the GATT obligations of contracting parties, see McGovern (1986, 30–31). The Section 22 waiver is discussed in McGovern (1986, chapter 14).

9. The waiver continued until the Uruguay Round, but the United States was forced to surrender it as part of the agreement of 15 December 1993.

The breadth of the waiver, coupled with the fact that the waiver was granted to the contracting party that was at one and the same time the world's largest trading nation and the most vocal proponent of freer international trade, constituted a grave blow to GATT's prestige. The waiver . . . was profoundly discouraging to many GATT supporters, and the United States was accused of hypocrisy.

The United States was hoist on its own petard when the special treatment for agriculture under the GATT rules formed the basis for the development, in the 1960s, of the common agricultural policy (CAP) by the EC. If trade in farm products had been subject to the same GATT disciplines as trade in manufactured products, the CAP could not have relied on variable import levies as its main protective device or on export refunds to subsidize the sale of its uncompetitive products on world markets. The United States has subsequently devoted considerable diplomatic effort to the aim of convincing the Europeans that the normal GATT disciplines should apply to agriculture.

The unsatisfactory nature of the GATT rules with respect to agriculture can be gauged to some extent by the frequency with which complaints on agricultural trade matters are brought to the GATT. Since the mid-1970s when the EC made the transition from being a net importer to being a net exporter of major agricultural products, the frequency of those complaints has increased. Of some thirty-two trade disputes brought before the GATT between 1976 and 1989, nineteen were agricultural trade complaints (Josling 1990, 157). Twelve of the agricultural trade panels involved complaints against the EC and included products ranging from pasta and wheat flour to poultry, citrus, beef, apples, sugar, and animal feed.

Previous Rounds of Negotiations

The GATT multilateral negotiating framework has achieved remarkable success in dismantling the high import barriers most industrial countries erected during the 1930s. Average tariff levels in the major industrial countries declined from about 40 percent in the mid-1940s to less than 5 percent at the completion of the Tokyo Round in 1979 (Baldwin 1987). The major reduction in protection experienced for manufactured products did not occur for agricultural products because most agricultural protection was by means other than tariffs.

Since the EC was formed there have been three GATT negotiating rounds held prior to the Uruguay Round: the Dillon Round (1960–62), the Kennedy Round (1964–67), and the Tokyo Round (1973–79). Agricultural negotiations formed part of all three rounds but domestic support measures were not included. Not surprisingly, the progress made in reducing agricultural protection was meager. The main features of the rounds as they relate to EC agricultural trade are discussed in the following sections.[10]

The Dillon Round

The focus of the discussions in the Dillon Round was on negotiating compensation for traditional exporters for loss of markets following the creation of the EEC. The case for compensation was based on the expectation that the imposition of a common external tariff (details of which were unavailable when the Dillon Round commenced in September 1960) would reduce the level of imports. An added complication was the undecided nature of the CAP at the time of the Dillon Round. The round essentially comprised two parts: a compensatory phase conducted under Article XXIV:6 and a reciprocal phase of tariff reductions conducted under Article XXVIII[bis].[11]

As far as agriculture was concerned, the move to a common external tariff and the proposal to change from import duties to variable levies for some products violated previous tariff bindings (Warley 1976, 379). In the course of the negotiations on compensation, the EC agreed to enter into GATT bindings at low or zero rates of duty for oilseeds, oilseed meals, and manioc (also known as cassava or tapioca). These products were relatively unimportant at the time but, as we discuss in chapter 3, these bindings provided the basis for significant growth in EC imports of "cereal substitutes," which was a source of much aggravation to the EC farmers and policymakers. In addition, the EC agreed to a 20 percent ad valorem tariff for sheep meat.

However, the EC denounced the tariff bindings on products that were to be covered by the CAP. The likely impact on U.S. exports to the EC was alarming, and the United States consequently sought guaranteed access to the EC market at its then existing export levels for products affected by the CAP. The U.S. claim for compensation was rejected by the EC. As Warley (1976, 379) noted, "completing the Dillon Round and thereby not impeding the progress of the Community was judged to be more important than resolving the agricultural issue," and the United

---

10. For further discussion, see Evans (1971, chapter 12), Curzon and Curzon (1976, chapter 2); Warley (1976, chapter 4); and Winham (1986, 146–58).

11. See Curzon and Curzon (1976, chapter 4), for further discussion.

States and the EC formally agreed that the United States had "unsatisfied negotiating rights." These were embodied in "standstill agreements" for wheat, rice, maize, sorghum, and poultry that recognized the
rights held by the United States on 1 September 1960.

The Dillon Round outcome, whereby the United States and the EC
agreed to recognize the unsatisfied negotiating rights of the United
States, permitted the establishment of the CAP and its variable levy
system. By so agreeing, the United States and other exporters lost the
opportunity to influence the mechanisms of the CAP in the future. The
standstill agreements were soon to be tested with the implementation of
the EC sluice gate and levy system for poultry imports in July 1962. The
EC policy on poultry threatened the growing U.S. exports of broilers to
Europe and in particular to West Germany. The effect of the EC poultry
regime was to triple the level of protection provided to the poultry
industry, and the United States claimed compensation on the basis of its
"unsatisfied negotiating rights recognised by the Dillon Round" (Curzon
and Curzon 1976, 211). The United States attempted to obtain modifications to the EC's poultry regime, but, when diplomatic efforts failed, the
United States took the matter to the GATT. An independent GATT
panel found that the United States was entitled to compensation for its
loss of the West German broiler trade.[12] As a result, the United States
withdrew concessions on potato starch, light trucks, brandy, and dextrin
amounting to approximately $26 million worth of imports from the EC.
As noted by the Curzons (1976, 212–13), these products were chosen to
impact almost exclusively on EC exports: "Thus Volkswagen buses and
French cognac were to be penalized in the U.S. market to compensate
for the loss of the German broiler market by American producers." But
the importance of the "chicken war," as it became known, went far
beyond the value of trade involved. The United States was testing the
interpretation of its unsatisfied negotiating rights and signaling its intention to take a hard line on agriculture in future negotiations, "lest the
development of the CAP resulted in other agricultural exports suffering
the same experience as its broilers" (Warley 1976, 381).

The Kennedy Round

In the lead-up to the next negotiating round, the United States made it
clear that, as the world's largest exporter of agricultural products, it

---

12. As determined by the GATT panel, the value of trade affected was $26 million.
See Curzon and Curzon (1976, 212), for details.

attached great importance to trade liberalization in agricultural products and to maintaining access to the market of the world's largest importer, the EC. Of particular concern was trade in cereals due to the importance of cereals in agricultural trade and the U.S. perception that European agricultural policy would have a major impact on the cereals sector.

During the Kennedy Round, a suitable framework for agricultural trade was discussed at length, with the United States arguing for greater reliance on market-oriented forces and the EC favoring a system of managed markets via international commodity agreements. In this round, the philosophical difference between the U.S. and EC positions proved to be irreconcilable and has continued to surface in subsequent rounds of negotiations. Progress with the negotiations was limited, partly because of the widespread use of nontariff barriers for agricultural trade and partly because the stage of development of the CAP made it difficult for the EC to agree on a negotiating mandate. The EC support regimes for dairy products, beef, sugar, and fruit and vegetables were not finalized until July 1966, some two years after the commencement of the round.

The Community's negotiating mandate on agricultural products consisted of two related elements: a method for measuring the margin of protection or support (better known by its original French name of *montant de soutien*) provided by each country to its agricultural producers, and a proposal to bind the margins of support at existing levels (Evans 1971, 209). The EC defined the margin of support as the difference between the price of the product on the international market and the actual remuneration received by the producer. This conceptually simple measure provided a means for encompassing the effects of the vast array of nontariff barriers used to protect the agricultural sector, but its implementation presented serious difficulties. The determination of the international price of a good was not straightforward because prevailing prices were influenced by the effects of protectionist measures. The matter was further complicated by the existence of a wide variety of grades and quality differences for agricultural products. As discussions progressed, it became apparent both that the EC proposal was *not* to negotiate the margin of support but rather to maintain the status quo and further that the status quo to be maintained was not the margin of support but rather the level of remuneration received by producers. Evans's view (1971, 211) was that "The margin of support was simply another name for a variable levy," but the proposal would have determined the maximum amount of support that could be applied. The U.S. delegation objected to the proposal on the grounds that it would extend the use of the variable levy system by the EC beyond those products for which it was then employed, including the products for which tariffs were bound in the Dillon Round. However, a major problem with the United States accepting the *montant de soutien* proposal was that it would have subjected U.S. domestic agricultural policy to international scrutiny, something that the United States was unwilling to accept. The U.S. delegation, recognizing that the variable levy system could not be eliminated, continued to press its claim for a stated share of the EC market. As noted by Evans (1971, 212), "Some of the heat generated by this debate resulted from the failure of either side to state its proposals with precision."

In the cereals negotiating group, the EC continued to press for its concept of managed markets that involved equilibrium between production and consumption, together with stable prices. This was to be achieved via the binding of each participant's *montant de soutien*. The major grain-exporting countries (the United States, Canada, Australia, and Argentina) advocated an arrangement that would at least maintain their current market access to the major importing countries. Agreement was not reached because the self-sufficiency ratios demanded by the EC and the United Kingdom (90 percent and 75 percent, respectively) were seen to be excessive by the United States and the other exporters. However, agreement was reached on the need to share the burden of food aid more equitably among donor countries, with the result that a Food Aid Convention was incorporated into the International Grains Agreement negotiated by the cereals group. The 1967 International Grains Agreement, like the earlier wheat agreements that it replaced, aimed at raising prices for wheat by means of multilateral commitments by exporters and importers. In the event, the new agreement proved ineffectual because wheat prices fell below the floor price in 1968.

As with the previous negotiating rounds, the benefits to agricultural trade were of far less significance than the benefits to manufacturing. The disappointing results with respect to agriculture can be attributed to two main factors: namely, the round did not address nontariff barriers, and both the EC and the United States rebuffed any proposals that would have subjected their domestic agricultural policies to international scrutiny.

The Tokyo Round

The next round of multilateral negotiations, which has become known as the Tokyo Round, opened in September 1973 amid growing concern over the impact of the "new protectionism" that was affecting not only agricultural markets but industrial markets as well. The oil crisis of the early 1970s, combined with difficulties in the international monetary system, instability in commodity markets, rising inflation, and unemployment led to concern that the long period of postwar economic expansion and prosperity was at an end.[13] The traditional trade balance between the major trading powers of North America and Europe was being challenged—initially by Japan, then by a number of other newly industrialized countries. Faced with increased competition (even on their home markets) and a stagnant world economy, it is not surprising that governments resorted to more protectionist and interventionist measures in an attempt to allay nationalistic concerns. As Golt (1978, 7) noted, "it is by now impossible to ignore that, alongside the avowed protestation by governments of their devotion to the open and liberal trading systems, essentially protectionist actions have been increasing on both sides of the Atlantic." Against this background, it was perhaps inevitable that the Tokyo Round was concerned with reducing or eliminating tariffs, nontariff barriers, and other measures that impeded or reduced trade in both industrial and agricultural products (including tropical products), as well as with strengthening the GATT codes of conduct.

Agriculture was again a key element in the negotiations, as it had been in the Kennedy Round. The Tokyo Round negotiations made very little headway during the years 1974–77, mainly due to the impasse in the agricultural negotiations. The United States and the EC entered the negotiations with positions fundamentally unchanged from the previous round. The EC continued to maintain that "the CAP was inviolable, and that neither its principles nor its mechanisms could be subject to negotiation in the Tokyo Round" (Winham 1986, 156). EC negotiators further insisted that a separate negotiating group be established for agriculture because of the special characteristics of the agricultural sector, a demand eventually accepted by the Americans. In addition, the EC continued to press for international commodity agreements—which both raised and stabilized prices—as a means of managing international commodity trade. If successful, such agreements would have had the effect of reducing the cost of export subsidies and the political exposure of the CAP (Harris, Swinbank, and Wilkinson 1983, 278).

The U.S. negotiating position differed from the EC's on all the major issues. The United States wanted to negotiate agriculture identically with other products, which would have meant subjecting agricultural trade to the same disciplines as those applying to industrial goods.

---

13. See Golt (1978, 5–9), and Winham (1986, 3–9), for elaboration on this point.

The U.S. government sought to eliminate the use of export subsidies in agriculture, an objective that was diametrically opposed to one of the major mechanisms of the CAP. In addition, the United States sought to expand the world market for agricultural goods and was specifically opposed to the concept of managed and stabilized markets inherent in the commodity agreements being proposed by the EC (Winham 1986, 156–57).

The negotiations got bogged down by what was ostensibly a procedural matter—the proper framework for the agricultural negotiations—but was, in reality, much more significant. The EC insisted that the negotiations with respect to agricultural trade should remain separate from the remainder of the negotiations, a logical position given the EC's mandate to avoid any negotiation of the CAP. The United States was equally insistent that agriculture should be part of the broader negotiations to provide an opportunity to trade off gains between sectors. The impasse was broken by the changes in key players that occurred following the election of Jimmy Carter in November 1976. His Democratic administration was less concerned with EC agricultural subsidies (perhaps because traditionally the Democrats have taken a more interventionist approach), and more concerned with supporting multilateral trade initiatives, than were the Republicans.[14] In July 1977, the new special trade representative, Robert Strauss, met with the EC negotiators in Brussels and developed a timetable for the conclusion of the Tokyo Round, including negotiations on agriculture.[15] With the impasse broken, the negotiations resumed, and there was soon progress in all areas, including agriculture. The agricultural negotiations were handled by a separate group, with different negotiating procedures from those applying to industrial products. Discussions in three major product areas—grains, bovine meat, and dairy products—were handled by separate subgroups.[16]

The Tokyo Round was concluded in April 1979. The final act of the round occurred at a meeting of the contracting parties in November 1979 at which the agreements reached during the round were integrated into

---

14. As Winham (1986, 165), notes: "It is difficult to assess the importance of this factor, but it may have been decisive. For example, one official in the U.S. trade bureaucracy expressed the view that had Gerald Ford won the presidential election in 1976 there probably would not have been a Tokyo Round agreement."

15. For further discussion of the change in the U.S. position, see Winham (1986, 164–67).

16. For discussion of the work in these subgroups, see GATT (1979, 24–34).

the GATT.[17] For agriculture, the results of the Tokyo Round were as limited as those achieved in the Kennedy Round. Agreement was reached on tariff reductions and on binding some tariffs at zero levels (for example, peanuts by the EC and soybeans by Japan), relaxing some quantitative restrictions (particularly by Japan), and concluding two commodity agreements for meat and dairy products (neither of which included market-stabilization arrangements). In addition, agreement was reached on establishing a consultative council to improve trade in agricultural products.

One outcome of the Tokyo Round with the potential to improve the conduct of agricultural trade was the Code on Subsidies and Countervailing Duties. The code obliged signatories to avoid causing harm or prejudice to the interests of other parties through the use of subsidies, prohibited the use of export subsidies for manufactured goods, and attempted to define further the meaning of "equitable share of world trade" for export subsidies on primary products.[18] However, it is widely acknowledged that the implementation of the code through challenges within the GATT system on the use of certain practices (e.g., the EC's use of export refunds) has not been successful. Not surprisingly, the definition of concepts such as "equitable share" has provided ample scope for debate and has tended to negate the purpose of the original agreement.[19]

As with the previous rounds, the Tokyo Round failed to deal with two crucial issues: domestic agricultural support policy and the special treatment of agriculture under the GATT rules. It was clear at the end of the negotiations that the EC and Japan were unwilling, for reasons of food security and domestic policy, to significantly expand imports of agricultural goods. In addition, despite attempts to improve the conduct of trade, the special treatment for agriculture under the GATT rules remained fundamentally unchanged. These issues would be raised again in the next round of negotiations.

---

17. For summaries of the agreements, see GATT (1979).
18. See Winham (1986, 422–24), for a summary of the code.
19. For discussion of the experience with the code, see Hathaway (1987, chapter 5), and Hartwig, Josling, and Tangermann (1989, chapter 3).

CHAPTER 2

World Agriculture in Disarray

In 1973, the first edition of D. Gale Johnson's seminal work on agricultural trade policy, *World Agriculture in Disarray,* was published. In his opening argument, Johnson (1973, 1) states that

> Products from the land are being produced at high cost in some parts of the world while elsewhere farm products that can be produced at low cost cannot be sold at all or only with great difficulty. The prices of farm products are manipulated by most governments and without any real knowledge of the consequences of such manipulation. In some countries consumers are forced to pay extremely high prices for many items on their food bill when comparable products could be made available at much lower prices. Economic relations among many friendly countries are soured by rigid adherence to economically unjustified restrictions upon trade in farm products.

A decade and a half later, as the Uruguay Round got underway, Johnson (1987, 142) posited: "Unfortunately, it takes little more than casual observation to permit one to conclude that the disarray has not diminished in the intervening period. It is apparent that the disarray has, in fact, deepened."

The purpose of this chapter is to provide an overview of the state of world agriculture in the early to mid-1980s and to review the major causes of the crisis.

Evidence of the Disarray

The disarray in world agriculture was manifested by sharp falls in international commodity prices and farm incomes, burgeoning stocks of surplus commodities, increased burdens on consumers and taxpayers, and heightened tension between major trading nations. As world commodity prices plummeted, farmers experienced severe financial stress, particularly in those countries where they relied on world markets for their incomes. Even in countries such as the United States, where farm support programs provided subsidies for producers, farmers were experiencing financial difficulties, land values were falling, and farm bankruptcies were increasing dramatically during the 1980s. In the EC, where farmers were insulated from the effects of falling world prices, the cost of support escalated. The effects of governments' distortionary agricultural policies were not confined to the rural sector: the problems flowed through to the rest of the economy, affecting employment, investment, and exchange rates.

Alongside the agricultural surpluses and excess productive capacity in much of the industrial world, widespread malnutrition persisted in many parts of the developing world, with serious famine occurring in some isolated areas. In many developing countries, economic policies discriminate against agriculture by maintaining internal food prices at uneconomic levels and by taxing the agricultural sector to support the development of the manufacturing sector. The position of the agricultural sector in the developing countries was disadvantaged further by the agricultural trade policies followed by the industrial countries.

Falling Commodity Prices in the Mid-1980s

World prices for the major agricultural commodities peaked in the early 1970s and again, to a lesser extent, in the early 1980s. By the mid-1980s, world prices for wheat and other grains, sugar, dairy products, and meat were falling in nominal terms and were well below their earlier peaks in real terms.[1] For wheat, the most important agricultural commodity in world trade, nominal prices rose sharply in the 1970s but declined during the early 1980s (see fig. 2.1). Indeed, by 1986 prices in real terms were only one-third of their 1974 value (Hathaway 1987, 22). In addition, the volatility of world wheat prices has been estimated to be "up to 50 percent greater than it would have been in the absence of restrictive trade policies and enormous agricultural subsidies by industrial countries" (Miller 1987, 12).

For sugar, an important commodity for many developing countries, world production and consumption grew rapidly during the 1970s, with immense swings in prices throughout the decade. During the 1980s, production continued to increase, even though consumption had slowed, with the result that stocks rose and world prices plummeted. In 1985, the real price of sugar was 7 percent of its 1974 peak and only 11 percent of its next peak in 1980 (Hathaway 1987, 39). The world sugar market has long been recognized as exhibiting a high degree of price volatility, but that

---

1. See Hathaway (1987, chapter 3), for a detailed discussion of the pattern of trade and the price trends for the various commodities.

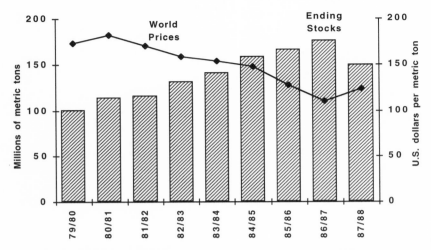

Figure 2.1. World wheat stocks and prices, 1979–80 to 1987–88. (World price is U.S. No. 2 HRW, Gulf ports. Data from International Wheat Council [1991, table 1].)

inherent variability has been increased by about 20 percent as a result of domestic sugar policies (Miller 1987, 12).

In countries that maintained relatively stable domestic support prices, the effect of falling world prices was to increase the "price adjustment gap." This was defined by Miller (1987, 27–28) as the ratio of the internal support prices to the current world price for the commodity. Although somewhat crude,[2] the adjustment gap indicates the reduction in farm support prices needed to bring domestic prices into line with world prices. In 1987, the ratio for the U.S. target price for wheat was approximately 1.5; for EC wheat, the ratio was approximately 2; and for Japanese rice, the ratio of the producer price to the world rice price was approximately 8. The Japanese ratios for sugar and butter were of a similar magnitude but were somewhat lower for beef, at approximately 2.5. While the U.S. and EC ratios for sugar and butter were lower than the Japanese, their magnitudes were nonetheless such that a significant adjustment would have been required to bring prices into line with world market trends.[3]

---

2. It must be remembered that the internal prices actually received by farmers may be below the support prices, and world prices would rise if domestic support measures were abolished.

3. See Miller (1987, charts IV.2–IV.5), for details.

Falling commodity prices combined with shrinking markets resulted in declining agricultural export earnings during the 1980s for many countries, from Argentina and Australia to Canada, Thailand, and the United States. As we discuss later in this chapter, the crisis in agricultural markets together with the increased use of export subsidies resulted in increased international tension and calls for reform of the GATT rules for agriculture.

Falling Farm Incomes and Land Values

Not surprisingly, the crisis in commodity markets resulted in many farmers experiencing severe financial hardship—particularly those in countries like Australia that did not support farm incomes to any significant extent. Farm incomes in Australia had recovered briefly following the drought in the early 1980s but slumped again in 1985–86 due to the decline in world prices. Wheat farmers were particularly hard hit. Reduced farm incomes together with a lack of confidence in the future conduct of the world wheat market were reflected in falling land values. Australia's Bureau of Agricultural Economics estimated that over the period of June 1985 to June 1987 land values in the broadacre sector of Australian agriculture declined on average by 25 percent in real terms (Kingma 1987). Marked differences were observed between industries and between states. The largest declines were 47 percent in real terms for wheat–sheep farms in Western Australia and the eastern states. Rising debt levels, higher interest rates, and falling commodity prices led to foreclosures, bankruptcies, and an increased rate of exodus from farming.

Many American farmers were also under financial pressure in the mid-1980s, despite the existence of government support programs. A study by Tweeten (1986) shows that rural land values across the United States fell on average by 17 percent in nominal terms between February 1981 and April 1985, but in some states the declines were much greater, with Iowa experiencing a 47 percent drop. Adjusted for inflation, real land values in the Corn Belt had fallen to less than half their 1981 values by April 1985. Tweeten attributes the dramatic fall in land values to a number of factors: high interest rates and falling rents; declining U.S. exports and the accumulation of large commodity stocks; efforts to reduce government deficits including farm program spending; and uncertainty about the new farm legislation to be enacted in 1985. The support available to farmers through the commodity programs failed to offset the negative impact of the macroeconomic factors. The introduction of the 1985 Farm Bill would initially depress farm earnings and land values even further (Tweeten 1986, 29). The effects on rural communities of the collapse in land values, the resultant farm bankruptcies, and the failure of agricultural banks and rural businesses are well known.

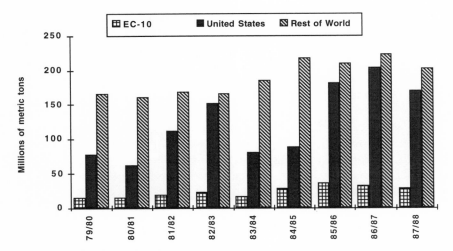

Figure 2.2. World cereals stocks by region, 1979–80 to 1987–88. (Data from FAO, *Food Outlook*, table A.8, various issues.)

European farmers were, by and large, insulated from the world agricultural crisis in the mid-1980s by the nature of the CAP, and rural land values remained buoyant. With the introduction of milk quotas in the dairy industry in 1984, quotas themselves acquired value at the expense of other assets including land.

Burgeoning Stocks

Despite the severe price drops being experienced by all the major commodities in international trade, stocks continued to increase during the 1980s. As shown in figure 2.1, world wheat stocks rose approximately 70 percent between 1980–81 and 1986–87, and greater increases of approximately 250 percent and 140 percent, respectively, were experienced for rice and coarse grains. World stocks for all cereals approximately doubled during the period 1980–81 to 1986–87, but by far the greatest increase occurred in the United States where cereals stocks more than tripled (see fig. 2.2). During that period, world cereals stocks as a percentage of world consumption rose from approximately 16 to 27 percent. The policies that resulted in the EC changing from a net importer of cereals to a net exporter at the end of the 1970s and the subsequent increase in stocks during the 1980s are explained in the next chapter.

In the EC, cereals stocks doubled during the period 1980–81 to

1986–87 while butter and beef stocks increased at an even more dramatic rate. EC beef stocks rose to over seven hundred thousand metric tons—approximately 30 percent of world trade in beef (Miller 1987, 12).

Rising Costs to Taxpayers, Consumers, and the Economy

The policies of the major industrial countries of insulating their domestic producers from the falling world prices meant that, not only did the levels of protection increase, but the costs imposed on taxpayers, consumers, and the wider economy also rose. One factor that had changed since Johnson (1973) first drew attention to the link between farm policies and the disarray in world agriculture was the effort being devoted to document and measure the costs of protection by organizations such as the World Bank, the U.S. Department of Agriculture (USDA), and the Organisation for Economic Co-operation and Development (OECD).[4] In addition, an extensive array of studies on the effects of protection was undertaken by the Australian Bureau of Agricultural and Resource Economics (formerly the Bureau of Agricultural Economics) in an attempt to stimulate the process of international agricultural policy reform.[5]

As stocks grew in the EC, so too did the costs of disposing of the surplus on world markets. The increased burden of the export subsidies was such that in 1984 the EC was forced to increase the maximum amount of the member states' value-added tax (VAT) contributions to the EC budget from 1 percentage point to 1.4 percentage points, effective from 1 January 1986. The escalating budgetary costs of the CAP, the creative accounting required to keep agricultural spending within the EC budget, and the losses incurred by disposal of the stocks (in particular, butter), are discussed at the beginning of chapter 5. In the United States, a major change in policy saw the introduction of the Export Enhancement Program as part of the 1985 U.S. Farm Bill. As a result, the annual cost to taxpayers of the U.S. farm programs exploded from between $3–5 billion in the early 1980s to $17 billion in 1985 and $30 billion in 1986 (Miller 1987, 12). Some indication of the magnitude of the annual transfer costs imposed on consumers and taxpayers by the support policies of the major industrial countries is provided by Miller's

---

4. See, for example, World Bank (1986), USDA (1987; 1988a; 1988b), and OECD (1988).

5. See BAE (1981; 1985), Stoeckel (1985), ABARE (1988), Roberts et al. (1989), and Riethmuller et al. (1990).

(1987, 12–13) estimate of nearly $700 per nonfarm family in the United States and more than $900 per nonfarm family in the EC.

Studies undertaken by the OECD (1988) and USDA (1988a; 1988b) show that the levels of protection afforded to producers in the major industrial countries increased markedly from the early to mid-1980s. These studies used the concept of producer and consumer subsidy equivalents (PSEs/CSEs), which aims to summarize the effects of a wide variety of government policies into one parameter and allows comparisons to be made across countries, commodity markets, and types of policies. Such comparisons would otherwise be impossible. The PSE is the level of producer subsidy needed to replace farm programs and leave farm incomes unchanged; the CSE is defined correspondingly.[6] The range of policies considered when calculating PSEs was initially commodity-specific and included pricing policies, deficiency payments, input subsidies, storage subsidies, and transport subsidies. However, work by the OECD (1987) broadened the policy coverage to include indirect income support and government programs such as structural programs, research, and extension, which are not commodity-specific.[7] PSEs are often expressed as the total value of subsidies as a percentage of adjusted producer income (cash receipts plus net direct payments). This is the form in which the PSEs calculated by the OECD are shown in table 2.1.

It is clear from table 2.1 that Japanese agriculture received the highest levels of protection in all categories. During the 1980s, there was a general increase in protection and, apart from New Zealand, the increase in crop protection was greater than for livestock. The rapid increase in levels of protection for crop products reflected the plummeting cereals prices as well as the domestic price policies aimed at maintaining produce prices at relatively stable levels. In the case of the EC, the ecu–dollar exchange rate has also had an important impact on the PSEs. As shown in figure 2.3, the U.S. dollar depreciated against the ecu, beginning in 1985 and continuing through early 1988. With world commodity prices denominated in U.S. dollars, the falling dollar meant that the "price gap" between the EC price and the world price expressed in ecu increased, resulting in higher PSEs. The falling dollar also increased the budgetary cost of disposing of surpluses on the export market. As noted by the USDA (1988b, 106), "The calculation of PSEs . . . has made the extent of subsidies to agriculture more transparent to commodity groups, policymakers and the pub-

---

6. The methodology was developed by Josling in an FAO (1973) study.

7. Work by the USDA (1988a) extended the range of measures to include the effects of exchange rate distortions in several developing countries.

TABLE 2.1. Net Percentage PSEs by Commodity and Country, 1979–86

	1979	1980	1981	1982	1983	1984	1985	1986
Australia								
Crops	3.4	4.8	8.2	14.7	8.2	8.6	12.7	18.4
Livestock								
products	10.8	11.7	11.3	17.3	17.2	15.3	16.7	14.1
All products	8.1	9.4	10.2	16.6	13.5	12.8	15.3	15.3
Austria								
Crops	37.7	23.3	11.3	21.0	20.8	11.6	16.5	n.c.[b]
Livestock								
products	37.0	40.4	46.8	43.3	48.7	47.0	41.8	n.c.
All products	37.2	35.1	36.8	36.3	40.6	36.5	34.1	n.c.
Canada								
Crops	13.7	14.8	16.2	19.7	19.2	25.2	39.7	53.5
Livestock								
products	31.2	31.0	30.4	30.9	35.7	37.4	39.5	39.3
All products	23.7	23.5	23.5	25.8	27.7	31.9	39.6	45.7
EC–10								
Crops	41.2	18.2	23.6	32.1	25.6	17.9	37.0	60.5
Livestock								
products	45.3	42.7	34.5	32.8	37.1	36.7	40.7	45.4
All products	44.3	36.4	31.7	32.6	34.2	31.4	39.7	49.3
Japan								
Crops	77.7	70.0	64.4	76.7	79.4	80.8	84.5	92.0
Livestock								
products	44.8	36.0	39.6	38.4	43.7	43.5	42.7	52.3
All products	64.3	54.3	53.1	59.4	63.3	64.9	66.7	75.0
New Zealand								
Crops	1.8	3.6	10.4	13.1	7.8	8.5	9.7	14.6
Livestock								
products	15.3	16.2	23.4	28.0	36.6	17.7	19.6	31.8
All products	14.9	15.7	22.8	27.4	35.3	17.3	19.0	31.1
United States								
Crops	8.2	9.2	11.5	13.6	30.1	21.4	26.2	47.2
Livestock								
products	19.9	18.9	22.8	20.0	23.1	24.7	26.1	27.0
All products	14.7	14.5	17.7	17.1	26.5	23.3	26.1	35.4
Average[a]								
Crops	30.1	21.2	23.5	28.4	34.9	30.8	39.9	61.3
Livestock								
products	34.1	32.3	29.6	28.0	31.8	31.6	34.0	38.3
All products	32.6	28.1	27.3	28.1	33.1	31.3	36.3	47.0

Source: Organisation for Economic Co-operation and Development (1988, table III.1).

[a] Weighted average only of countries shown.

[b] n.c. = not calculated.

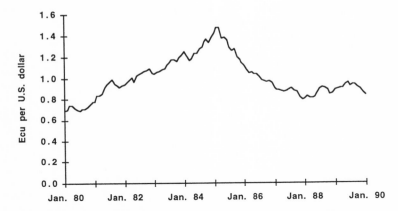

Figure 2.3. Ecu–U.S. dollar exchange rate, 1980–90. (Data from Commission of the European Communities, *European Economy, Supplement A,* various issues.)

lic." Clearly, domestic agricultural policies that resulted in nearly 50 percent of farm returns being provided in the form of subsidies—over 60 percent in the case of crop products—could not continue unchallenged.

In addition to the transfers and subsidies to producers, the farm support programs adopted in the United States, the EC, and Japan had a major impact on the national economies. The distortion of the trade structure resulting from the farm policies has an adverse effect on other sectors of the economy and reduces growth rates, trade in manufacturing, and employment. For example, a study of the EC undertaken by Stoeckel (1985) suggested that for the four largest countries (Germany, France, Italy, and the United Kingdom), agricultural support given for the period 1973–82 could have lowered manufacturing exports by 4 percent, increased imports by 5 percent, reduced manufacturing output by 1.5 percent, and resulted in a loss of about half a million jobs in the manufacturing sector. This is because, in addition to more resources in total being devoted to agricultural production in protected economies than would be the case without protection, total economic output is reduced as a result of fewer resources being allocated to less protected, and usually more efficient, industries.

Trade Conflict

In chapter 1 we noted the disproportionate number of trade disputes involving agriculture that have been brought before the GATT and the high frequency of complaints against the EC. Not surprisingly, the deteriorating trade environment of the 1980s led to increased trade tension. Although the EC remained the focus of a number of agricultural trade disputes (for example, beef, poultry meat, oilseeds, canned fruits, and citrus) the United States became the center of much of the trade conflict due to the introduction of the Export Enhancement Program in mid-1985. During the early 1980s, the strengthening U.S. dollar, combined with expansion of output by other exporters, led to a significant decline in the U.S. market share of world agricultural exports. For example, from 1980–81 to 1984–85, the U.S. share of the world wheat market fell from 42.5 percent to 39.9 percent (Argentina, Australia, and the EC increased their shares). For coarse grains, the U.S. share fell from 59.4 percent to 49.1 percent (Australia, China, and the EC increased their shares).

As we have already seen, this led to a massive accumulation of stocks and escalating budgetary costs for the United States. To reduce stocks and win back market share, the United States introduced generous export subsidies for targeted markets to counter what the United States claimed to be "unfair competition" from the other exporters, notably the EC.[8] The disposal of U.S. stocks was highly disruptive to world markets and depressed prices further. Grain farmers in other exporting countries petitioned their governments for compensation, and worldwide the cost of supporting agriculture increased.

The subsidized export sales resulted in strong protests from various governments. The Australian government threatened to remove U.S. bases from Australian territory, and the Argentines threatened to halt payment of foreign debt in retaliation. The protests were not restricted to grain exporters. Exporters of sugar, cotton, and dairy products protested at the U.S. use of export subsidies. The U.S. action was seen by the major agricultural exporters as the height of hypocrisy insofar as the United States advocated the freeing up of world trade. The American view was that the use of subsidies helped to sustain the international pressure on the Europeans to change their CAP and discontinue their export subsidies. By the late 1980s, the United States had regained much of its market share but at a high cost in terms of trade conflict.

---

8. For a discussion of factors influencing U.S. market share and stocks of wheat, see Roberts et al. (1989, chapters 2 and 3).

Causes of the Disarray

In the period since Johnson (1973) first drew attention to the relationship between domestic agricultural policies and trade, the adverse impact of national agricultural policies on world agricultural trade has become well understood. It is widely acknowledged that the principal factor leading to the disarray in world agriculture in the 1980s was the domestic agricultural policies followed in the major industrial countries. The oil crisis of the early 1970s and the dramatic increases in commodity prices that ensued provided the impetus for agricultural policies that encouraged investment and led to expansion of output.[9] As production outstripped consumption in the early 1980s, the insulation from world markets provided by most domestic agricultural policies—in particular those of the United States and the EC—meant that production continued unabated and, as we have seen, large stocks accumulated. With export markets stagnating and competition for those markets increasing, the food crisis of the 1970s turned into the trade crisis of the 1980s.

The adverse impact of national agricultural policies on the international marketplace in the early 1980s was exacerbated by a number of other interrelated factors leading to a significant change in the structure of agricultural trade, namely:

- a slowdown in world economic growth;
- high real interest rates, and the appreciation of the U.S. dollar;
- debt problems of developing countries; and
- changes in the agricultural policies of many developing countries and centrally planned economies.

The structure of agricultural trade in the early 1980s was significantly different from that which prevailed during the 1970s. Developed countries accounted for approximately two-thirds of world agricultural imports in the early 1970s. Their share fell from 67 percent in 1972 to 51 percent in 1980. During this period, the volume of world agricultural imports grew by some 34 percent. Most of the growth occurred in developing countries and to a lesser extent in centrally planned economies and was due to increased food consumption in those countries. The increased food imports were financed from oil revenues and foreign borrowings (Curran, Minnis, and Bakalor 1987, 90). The increase in

---

9. For an excellent review of national agricultural policies of both industrial and developing countries, see Hathaway (1987, chapter 4).

imports was met largely by increased exports from North America and the EC where, as discussed earlier, domestic agricultural policies were in place that tended to encourage production.[10]

The slowdown in world economic growth that occurred in the early 1980s flowed through to agricultural trade. In contrast to the strong growth in the volume of agricultural imports in the 1970s, the volume of imports grew by only 5 percent over the period 1980–84, and in 1982 the volume of trade for most agricultural commodities actually fell (Hathaway 1987, 16). Much of the slowdown in growth of agricultural imports in the early 1980s was due to the reduced demand for agricultural imports by the developing countries and the centrally planned economies.

Many developing countries had shifted from agricultural policies that taxed the farm sector to policies that favored agriculture or were neutral. This resulted in increased agricultural production and a reduction in imports. For example, in India substantial investment in infrastructure and input subsidies had largely eliminated imports of major food items by the early 1980s. In China, the liberalization of sections of its domestic economy resulted in substantial growth in the agricultural sector, in a reduction in wheat imports, and in the country becoming a major exporter of coarse grains.

Many developing countries were heavily dependent upon commodity exports and had high debt levels. The high real interest rates of the early 1980s and the strengthening dollar,[11] compounded by the fall in world commodity prices, exacerbated the debt problems these countries faced and reduced their capacity to import. In countries like Brazil and Argentina, as the debt crisis led to falling real exchange rates, agricultural production and exports were increased, adding fuel to the crisis in agricultural trade.

The high real interest rates of the time discouraged the holding of commodity stocks by the private sector, thus dampening demand. The move from fixed to floating exchange rates that occurred in many countries in the early 1970s and subsequent large movements in the relative values of currencies for the major agricultural trading countries added to the confusion in international markets. Although world commodity prices in U.S. dollars fell fairly steadily from 1980 to 1985, the effect of the strengthening U.S. dollar (see fig. 2.3) was that commodity prices

---

10. For a fuller discussion of the factors contributing to the change in the structure of world trade, see Tyers and Anderson (1992, chapter 1).

11. This is the opposite of the situation in the 1970s, when real interest rates were low or negative.

expressed in local currencies (such as pounds sterling or French francs) continued to improve during this period. At the peak of the exchange rate between the ecu and the dollar, the EC internal agricultural prices were only moderately above the world prices, and the cost of subsidizing EC exports was reduced without any change in internal prices. As we have already noted, once the dollar started to fall relative to the ecu in early 1985, the internal EC prices rose sharply relative to world prices, and the cost of subsidizing exports increased.

The deteriorating world trade environment, the increasing costs of supporting the agricultural sectors of the United States and other industrial countries, and the increasing awareness that—despite the escalating costs—most agricultural policies were failing to achieve their objectives all led to calls for reform. The crisis in agricultural trade provided a major impetus for embarking on the Uruguay Round, and the continued conflict in agricultural markets during the negotiations served as a reminder of the need for change.

CHAPTER 3

The CAP

For many years the CAP and the customs union, which had been established in the mid-1960s, were the main policy achievements of the EU.[1] Formulated in the 1960s, the CAP personified the EU both to its inhabitants and throughout the world. It swallowed up the major part of the EU's budget, its legislative provisions filled the pages of the *Official Journal of the European Communities,* and together with the budget, it was political battles over the CAP that dominated proceedings within the Council of Ministers.

CAP Objectives

The objectives of the CAP are set out in Article 39(1) of the EEC Treaty:

The objectives of the common agricultural policy shall be:

(a) to increase agricultural productivity by promoting technical progress and by ensuring the rational development of agricultural production and the optimum utilisation of the factors of production, in particular labour;
(b) thus to ensure a fair standard of living for the agricultural community, in particular by increasing the individual earnings of persons engaged in agriculture;
(c) to stabilise markets;
(d) to assure the availability of supplies;
(e) to ensure that supplies reach consumers at reasonable prices.[2]

---

1. The acronym *EU* is used throughout this chapter except when *EC* is more appropriate.
2. Article 39(1) EC as of 1 November 1993. But the former citations are retained in this chapter.

It is no exaggeration to claim that a farm income objective has dominated CAP policy making from the outset. This is despite the fact that it is far from clear what is meant by a "fair" standard of living, who is to be included in "the agricultural community," what levels of farm income are actually attained, or that item (b) above is really linked to item (a) by the word "thus," which implies that increased incomes are dependent upon increased productivity. Farming in the EU is incredibly diverse, embracing many different farm enterprises, farm business sizes, and land tenure arrangements, with major climatic and soil differences. Trying to measure farm incomes, or to define what is meant by "farm income," is not easy, particularly as there are many part-time farmers in the business. Farmworkers and landlords are almost certainly excluded politically from the concept of an "agricultural community," but many part-time farmers with only tenuous links with the farm sector seem to be included.

The policy objectives outlined in Article 39(1) are not ranked and from time to time may well be in conflict. Snyder (1985, 19) points out that this has been recognized by the European Court and that "Weighing the conflicting aims of the CAP is thus a political decision." Historically, consumer interests have been subservient to producer interests.

Article 39(1) does not list self-sufficiency as one of the CAP's objectives, although it does talk about assuring the availability of supplies. Food security is a complex notion, but distant memories of food shortages and television images of starvation around the world are emotive symbols that farm leaders can and do deploy to argue for a farm policy that delivers an overabundance of EU-produced food.

The need to avoid "desertification" by sustaining viable rural communities that are to be valued for their own sake because of the cultural and social values they embrace is also a crude, but politically effective, route for canvassing support for the farm sector. Over many centuries, much of the European environment has been shaped by farming and other rural land-use activities, and very few wilderness areas survive. When the Treaty of Rome was being negotiated, the dominant view was that environmental and farming concerns coincided and that farmers were the custodians of the countryside—though dissident views were expressed.[3] If Article 39(1) were to be redrafted today, the environmental pressure groups would undoubtedly demand that environmental concerns be explicitly listed among the policy's objectives.

---

3. Rachel Carson's *Silent Spring* had been published in 1963 for example.

A Brief History

The Treaty of Rome did not create the CAP. This was essentially a construct of the 1960s, reflecting in part the national policies it displaced and influenced by the deliberations of an intergovernmental conference held at Stresa in 1958. From the outset there were two strands of farm policy: "guidance" and "guarantee." Guidance, or structural, policy was concerned with reshaping the fragmented, unspecialized, and small farms that made up EU agriculture, while guarantee was a price support policy that, by supporting farm revenues, was supposed to be reflected in enhanced farm incomes. Broadly speaking, the thesis was that through "structural" policy EU farming would improve its efficiency and shed surplus labor, but during the adjustment process there needed to be price guarantees to sustain farm incomes.

The key decisions on the CAP's price support policy date from 14 January 1962 when, after a marathon session of the Council of Ministers, agreement was reached on the market support mechanisms that were to regulate six products (cereals, pig meat, poultry meat, eggs, fruits and vegetables, and wine). Although the price support regimes have been modified significantly on a number of occasions since, most notably in the introduction of the Mac Sharry "reforms" of 1992, the basic design of 1962 still applies. The 1962 decisions, it was subsequently claimed, established "three fundamental principles" on which the market and price support policies of the CAP were based.[4] They were the following:

- *market unity,* which allowed products to flow freely from one member state to another and implied that the levels of price support are common throughout the EU. However, as a result of monetary disturbances between 1969 and 1992, the use of special agricultural conversion rates and monetary compensatory amounts meant that this principle was not observed. Since 1 January 1993 a new agrimonetary system has been in place, as we will describe later in this chapter, but the reemergence of currency instability in the exchange rate mechanism (ERM) on 1 August 1993 placed this new system in jeopardy.
- *community preference,* which implied that third-country imports are only necessary if EU producers are unable to meet EU demand, is reflected in import controls that give a price advantage to local producers. As we will discuss later in this chapter, commu-

---

4. Commission (1975, par. 12).

nity preference was a characteristic feature of most CAP price support mechanisms, but it was under threat in the Uruguay Round of GATT negotiations.

• *financial solidarity,* which broadly speaking has meant that the full budgetary impact of the CAP's price support policies has been reflected in the EU's budget.

The 1960s were dominated by the development of the price support policy, with little progress made on the structural side. In 1968, the Mansholt Plan drew attention to the small and fragmented farms characteristic of EU agriculture as well as the emerging structural surpluses, particularly in the dairy sector, and it urged a major change in policy with an enhanced role for structural policy (Commission 1968). Nonetheless, it was not until the early 1970s that a limited set of structural policy measures was introduced.

The prospect of the enlargement of the EC in 1973 to embrace the food-importing United Kingdom, as well as a price boom on world commodity markets that led to talk of a world food crisis, basically gave a green light to CAP policymakers who then presided over an expansion in production associated with an increase in protection. There was no need to restructure EU agriculture if farm incomes could be supported by buoyant, if protected, prices and apparently unlimited opportunities for accumulation of surpluses at the taxpayer's expense. In spite of the fact that structural policy grew in relative importance from the mid-1970s, and its emphasis shifted from fostering commercial viability throughout the EU's farm sector to sustaining vulnerable farming communities in marginal areas, farm price support continued to dominate the CAP.

By the end of the 1970s, in the face of ever-increasing production coupled with stagnant consumption and depressed world markets, budgetary pressures began to force change in the CAP. However, creative accounting and a good deal of "fudge" ensured that the bankrupt CAP survived against the predictions of most of the pundits. Milk quotas, it is true, were introduced in 1984, and throughout the decade the Council and the Commission agonized over various ineffectual measures to curb the growth in production and expenditure. Agronomists, and others, increasingly talked about a "surplus" of farmland rather than a surplus of farmers, and in 1988 a limited set-aside scheme was introduced. In 1986, the EC embarked upon the Uruguay Round of GATT negotiations, a successful conclusion to which would imply some reform of the CAP. During the course of the negotiations, in January

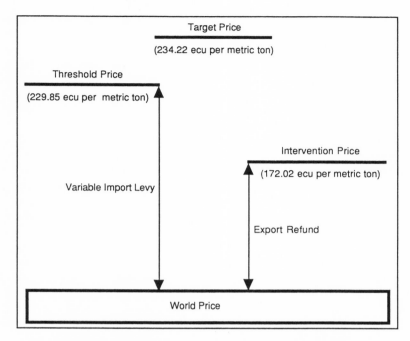

Figure 3.1. CAP price support for wheat (beginning of the 1990–91 marketing year)

1991, the then farm commissioner, Ray Mac Sharry, introduced a set of reform plans that, with some significant changes, were agreed to by the Farm Council in May 1992. For the arable sector, these switched part of the burden of support for farm revenues from consumers to taxpayers and for most farmers introduced a set-aside requirement if they were to receive area compensation payments. These changes will be outlined in chapter 5.

A Characterization of the CAP

For the major part of agricultural production of importance to EU farmers and the diet of EU consumers, the CAP has attempted to maintain farm-gate prices well in excess of the price levels that would obtain in the absence of the policy. Figure 3.1 describes the main elements of the market management policy mechanisms that applied to cereals. Similar policy mechanisms were in place for sugar, dairy products, eggs, and most meats. The Mac Sharry reforms of May 1992 modified the picture for cereals, and the GATT agreement of December 1993 abolished variable import levies. For the moment, however, we are concerned with the CAP as it was.

Three policy mechanisms are commonly used. First, imports are taxed. In the case of cereals, a variable import levy was determined daily, based upon the difference between the lowest observed world market price and a predetermined minimum import, or threshold, price. Second, exports are subsidized to enable EU traders to compete on world markets. These subsidies are known as export refunds (or, adopting French terminology, as export restitutions). Third, if market prices within the EU are weakening, some form of intervention is usually available that allows farmers, or more usually traders, to sell their product to intervention agencies at a price equal to, or somewhat less than, the intervention price.

Threshold prices were set higher than intervention prices to ensure "community preference," that is, the philosophy that EU produce should find an outlet on EU markets in preference to imported products. When the price gap is large, as it has been with wheat, the policy can induce severe problems. Thus, Canadian hard bread-making wheat was still imported into the United Kingdom, despite the country's overall surplus in wheat, because of its different quality characteristics. Consequently, millers actively sought alternative sources of protein, such as wheat gluten, to fortify their bread-making grists and minimize the price penalty incurred by utilizing third-country wheats.

The CAP typically forces a sizable wedge between market prices prevailing in the EU and the much lower prices on world markets. The "cost" of the policy to the EU's citizens is in two parts: higher food prices and the net cost to the EU's budget. The "invisible" transfer that the buyer pays to the seller in the form of higher prices can be significant but is typically difficult to determine with any degree of certainty. The problem centers on specifying the world market price. The taxpayer only becomes involved if the level of domestic production exceeds that of consumption and thus expenditure on intervention and export refunds exceeds the levy revenue generated on imports. But this soon became the norm. It should be remembered that the principle of financial solidarity applies. It is from the EU's budget that CAP price support payments are made, resulting in net budgetary transfers between the member states. Similarly, consumers in one member state may be paying inflated food prices for items produced by farmers in another member state.

Other CAP Price Policy Mechanisms

The CAP is not a uniform policy applied consistently throughout the farm and food sectors. There are some significant variations on the theme. These fall into two broad camps. First, in some product sectors, in an attempt to reduce either the political or the budgetary cost of the policy, the EU attempts to enlarge the market by making products available to selected users at a lower price than the full CAP intervention price. The EU market for butter fats is a particular case in point.

Second, for some important products such as tobacco and, until recently, oilseeds, the archetypal CAP regime described in figure 3.1 does not apply. Instead of a "managed market" system, pushing up market prices through a combination of import controls, intervention buying, and export subsidies, farm support is secured, in effect, by a deficiency payments scheme in which taxpayers rather than consumers meet the cost.

The oilseeds market is a particularly complex case because, arguably, EU policy induced a higher level of U.S. oilseed production than would otherwise have occurred. This was a consequence of the EU's cereals policy, which, in pushing up the cost of cereals in livestock feeds, encouraged a higher level of usage for other feed ingredients, or "cereal substitutes," to use CAP jargon. These often require fortification with protein; oilseed meal obtained as a joint product with vegetable oils from the crushing of oilseeds is one such protein source. Thus, large quantities of U.S.-grown soybeans are imported into the EU for crushing. The soybean meal is eagerly taken up by the animal feed industries, whereas the soybean oil has no ready outlet. As a result, on the back of its soybean imports, the EU is a major exporter of soybean oil to world markets, and vegetable oil prices in general are depressed. Oilseeds, as we shall see, played a central role in the Uruguay Round.

CAP and the Food Industries

The CAP has a profound impact on the food industries. For first-stage processors—those handling agricultural raw materials such as abattoirs, dairy processors, and sugar beet refiners—the policy has an impact on their volumes of production and hence upon the capacity utilization of their plant. Quotas, or other measures that limit the volume of raw material supplies, will have a clear impact on their profitability. Most CAP price support mechanisms also operate via these first-stage processed products. Thus, it is the prices of refined sugar and processed dairy products that are supported, not those of sugar beet or raw milk.

Consequently, these businesses are often as heavily implicated in the machinations of the CAP as are the farmers from whom they buy their raw materials and the storage companies that accommodate the intervention stocks.

Second-stage processors, such as biscuit manufacturers, take these first-stage processed products and transform them into more sophisticated consumer goods. The CAP raises their raw material costs, and they may have difficulty passing on these higher costs to the demanding retail stores. These so-called non-Annex II goods, because they are not listed as CAP products in Annex II of the Treaty of Rome, also attract variable import levies and export subsidies, reflecting their raw material components. The CAP thus adds to the managerial complexity of importing and exporting processed products and thus broadens the range of items potentially subject to fraudulent activities. As we saw in chapter 1, export subsidies on agricultural products have been tolerated in GATT, but on manufactured products they are not permitted.

Consequently, the European food industries were concerned when, in 1983, a GATT panel convened to consider the payment by the EC of export subsidies on pasta concluded that "durum wheat incorporated in pasta products could not be considered as a separate 'primary product' and that the export refunds paid to exporters of pasta products could not be considered to be paid on the export of durum wheat" (Harris 1994a, 197). The EC refused to accept this ruling and continued to pay export refunds on processed food products. But the food industries were clearly interested to see how this issue would be dealt with in the Uruguay Round.

CAP and the Budget

As we noted earlier, one of the central tenets of the CAP's price support policy is financial solidarity, which implies that all the financial consequences of the policy are reflected in the EU's budget. Thus, the revenues from import levies are paid into the budget, and expenditures on export refunds, intervention, and the like are charged against it. In contrast, structural policy has always been partially funded, with the member states meeting the remainder of the subsidy cost of activities undertaken on their territories. Although somewhat frayed at the edges after many years of political compromise and expediency, this sharp distinction between full funding of price policy and partial funding of structural policy remains essentially correct in the 1990s.

The concept of financial solidarity is linked with that of the unity of the market, and the so-called Rotterdam effect will often be cited as a relevant factor. When goods enter the EU they must pay customs duties, or import levies under the CAP, at the border, and they are then in free circulation in the EU. Thus, consumers in France or Germany may well be paying taxes that, in the case of goods imported through Rotterdam, will have been collected by the Dutch authorities. The rhetorical question then raised is, would it be fair for the Netherlands to benefit financially from the accident that the goods happened to be imported into the EU through a Dutch port? The response is emphatically no: such revenues should be treated as the own resources of the EU because they stem from the application of EU policies. Thus, customs duties and agricultural levies were from the outset treated as the EU's own resources.

For CAP price support on the expenditure side a similar argument applies. If there is to be a single market for farm products, then there has to be EU funding of the policy. The downside of this is that a moral hazard emerges. If a group of friends in a restaurant agrees to share the bill equally, whatever the individual's choice, no one will have an incentive to eat modestly because the exorbitant cost of choosing, say, lobster will be shared. If each individual follows the same strategy the bill will soar. Therefore, it is probably fair to claim that member states have not been as financially prudent with the CAP as they would have been with their own nationally funded policies. In particular, on an individual basis they will have sought to enact EU schemes with a specific benefit to themselves that they would probably not have pursued had the full budgetary burden fallen upon their own exchequers.

Green "Money"

The ecu (which was once known as the European Currency Unit) is made up of a basket of currencies of the member states, as shown in table 3.1. Following the devaluation of the Spanish peseta and Portuguese escudo on 13 May 1993, within the exchange rate mechanism (ERM) each currency had central rates against the ecu, as shown in the third column of table 3.1. (These central rates were unaffected by the decisions of 1 August 1993). Thus, on 13 May 1993, the percentage weight that each currency contributed to the ecu was as listed in column 4 of table 3.1. It can be noted that the German mark accounted for 32 percent of the value of the ecu, while the Greek drachma and the Portuguese escudo each represented less than 1 percent.

The ecu is used in all the EU's policies, and the budget is drawn up in ecu. The ecu forms part of the European Monetary System (EMS). Formally at least, it plays a central role in the ERM, and the Maastricht Treaty sees it as the fledgling currency for economic and monetary union. An ecu bond market exists, contracts are drawn up in ecu, and ecu bank accounts can be held.

CAP prices, as we saw in figure 3.1, are fixed in ecu. However, as we shall see later, appearances can be deceptive, and an important correction must be made to this simple statement.

Although CAP prices are expressed in ecu, to date all transactions are effected in terms of national currencies. This means that a conversion rate has to be applied to convert CAP prices such as import levies and export refunds into national currencies, and vice versa. If these conversion rates used in the CAP had always corresponded with the realities of international currency markets, then the CAP's agrimonetary complications would not have arisen. However, from an early date (1969) divergencies did appear, and the complications of green "money" and monetary compensatory amounts (MCAs) were introduced into the CAP.

In an attempt to maintain domestic price stability even though the ecu value of a country's currency was changing, all member states at one time or another used special conversion rates that differed significantly from the market exchange rate of the ecu, to convert from ecu to national currencies for all CAP-related transactions. These special conversion rates were known as green "money" or green conversion rates. This meant that CAP prices were not the same from one member state to another, despite the commitment to common pricing, and to sustain

TABLE 3.1.   The ecu on 13 May 1993

Currency	Units	Central Rate (1 ecu = ...)	Percentage Weight
Belgian/Luxembourg franc	3.431	40.2123	8.52
Danish kroner	0.1976	7.43679	2.66
German mark	0.6242	1.94964	32.02
Greek drachma	1.44	264.513[a]	0.54
Spanish peseta	6.885	154.250	4.46
French franc	1.332	6.53883	20.37
Irish punt	0.008552	0.808628	1.06
Italian lira	151.8	1,793.19[a]	8.47
Dutch florin	0.2198	2.19672	10.01
Portuguese escudo	1.393	192.845	0.72
British pound	0.08784	0.786749[a]	11.17
			100.00

Source: For the composition of the ecu see *Official Journal of the European Communities*, C241, 21 September 1989, p. 1.

[a] Nominal values.

these price differences between the member states border taxes and subsidies on intra-EU trade were required. These border taxes and subsidies were known as MCAs.

The 1992 Programme necessitated the removal of all barriers to intra-EU trade by 31 December 1992, and so a new green money policy was devised to apply from 1 January 1993.[5] However, this does not mean that the conversion rates then used in the CAP exactly reflected the value of the ecu determined on international currency markets.

The old system of green "money" and MCAs meant that some countries had undervalued green conversion rates—their support prices, above the nominally "common" price level, were sustained by "positive" MCAs—while others had overvalued green conversion rates with "negative" MCAs and price levels below the common level. The political objective was to move both sets of countries toward the common price level by "revaluing" undervalued and "devaluing" overvalued green conversion rates. The latter, unlike the former, was politically attractive to the farmers in the country concerned because it resulted in price increases in national currency terms. From Germany in particular, there was considerable resistance to the notion of removing positive MCAs by revaluing the green mark, while other countries, notably France, strongly objected to the price advantage German farmers were receiving.

A "solution" to this dilemma was devised in 1984. It is known as the "switchover" mechanism in that it switched positive MCAs into negative MCAs by redefining the common price level. The situation is illustrated in figure 3.2. At the time, the price gap determining the German MCA stood at +10.8 percent and that determining the French MCA at −5.9 percent.[6] A political decision was taken to reduce the German MCA gap by three percentage points. To do so, a "correcting factor" of 1.033651 (see table 3.2) was introduced into the MCA calculations so that the French MCA gap increased to −9.5 percent. Ecu prices remained unchanged, but in effect the ecu in which CAP prices were fixed was now worth 3.4 percent more than the ecu as defined in table 3.1. A "green ecu" had been born.

Subsequently, the switchover mechanism was used on a number of occasions, increasing the correcting factor as detailed in table 3.2 so as to, first, eliminate and, second, prevent the creation of positive MCAs for any currencies participating in the narrow band of the ERM.

The system was inflationary in that it pushed up the common level

---

5. For details, see Swinbank (1993b).
6. Measured as a percentage of the German and French price levels, respectively.

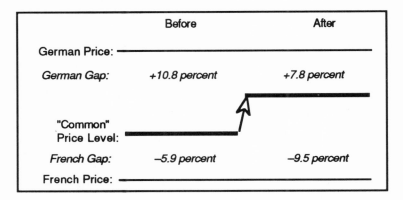

Figure 3.2. The 1984 price settlement and the switchover mechanism

of prices, and countries with negative MCAs were encouraged to devalue their green conversion rates to reestablish common CAP pricing throughout the EU. Common prices were in fact achieved on 1 January 1993 with the introduction of the new agrimonetary system. However, as table 3.2 shows, by this time the green ecu was worth 19.5 percent more than the ecu shown in table 3.1. Furthermore, the switchover mechanism was retained in the new system, and from May 1993 until its abolition in 1995 the correcting factor was 1.207509.

In the remainder of this book, it must be remembered that, in outlining CAP policy mechanisms, all CAP prices are set in the green ecu, which was worth over 20 percent more than the ecu more commonly encountered (and which the European Commission often referred to as the commercial ecu).[7] However, in the context of the GATT negotiations and the Uruguay Round agreement, the EU entered into binding commitments expressed in terms of commercial ecu. The divergence between commercial and green ecu, and a threat that the divergence might widen in future years, posed a policy dilemma for the EU as it strove to render the CAP compatible with the GATT agreement. We must return to this theme later in the text.

The CAP and World Markets

The CAP has been widely criticized. It pushes up food prices and bears particularly heavily upon low-income households. Significant intra-EU

---

7. In other texts, Swinbank has frequently used the term *real ecu*.

financial transfers are generated from member states that are net food importing to those that are net food exporting. In pushing up the raw material costs of the food industries, the CAP embroils them in the machinations of policy and causes them to devote considerable management time and effort to either, at best, coping with the policy or, at worst, exploiting it. It encourages fraud and corruption, and its budgetary costs lead to tensions between the member states and may crowd out other spending priorities from the EU's agenda. Using intervention as a major policy tool leads to the buildup of unwanted stocks of food, which are a political embarrassment and a further waste of the scarce resources needed to store the stocks. The CAP's untargeted approach is inequitable in that it showers largess on the larger, prosperous farmers and the productive farming regions, while small farmers and disadvantaged re-

TABLE 3.2.   The Correcting Factor

Date	Correcting Factor	Comment[a]
1 April 1984	1.033651	Introduction of system, reduction of 3 percentage points in German monetary compensation amounts (MCAs)
22 July 1985	1.035239	Devaluation of Italian lira
8 April 1986	1.083682	Revaluation of German mark and Dutch guilder
4 August 1986	1.097805	Devaluation of Irish punt
15 January 1987	1.125696	Revaluation of German mark and other currencies
1 July 1987	1.137282	1987–88 CAP price fixing, reduction of 1 percentage point in German and Dutch MCAs
6 January 1990[b]	1.145109	Devaluation of Italian lira on its entry into the narrow exchange rate mechanism (ERM) band
14 September 1992[b]	1.154338	Devaluation of Italian lira
17 September 1992[b]	1.157346	Devaluation of Spanish peseta
23 November 1992[b]	1.195066	Further realignments within the ERM and in particular the fixing of a new nominal central rate for sterling
1 February 1993[b]	1.205454	Devaluation of the Irish punt
14 May 1993[b]	1.207509	Devaluation of Spanish peseta and Portuguese escudo

[a] Each reshuffle within EMS involves the fixing of a new rate for non-ERM currencies.

[b] And ecu support prices then reduced to claw back 25 percent of the price rise resulting from the increase in the correcting factor.

gions tend to be ignored. Price support as a mechanism for income support is inefficient in that it encourages unwanted production. Some of the benefit will be captured by the input supply industries rather than the farm sector. Much of the benefit for the farm sector will be captured by landowners rather than tenant farmers. High land prices will result in more marginal land being drawn into cultivation and the application of more fertilizers and agrochemicals to increase yields, causing environmental damage. In short, the CAP generates major economic distortions within the EU.

But the CAP's adverse impact is not confined to the EU. The EU is a significant importer and exporter of agricultural and food products, and consequently its domestic support policies are bound to have an effect on world markets. Thus, it is no surprise that the CAP is universally denigrated by the EU's trading partners, that disputes about the CAP have soured U.S.–EU trade relations, and that the CAP was one of the contentious issues at the heart of the Uruguay Round of GATT negotiations. The CAP has the potential to affect the level of world market prices, their stability, and the price relationships between different farm commodities.

By expanding farm production in the EU and depressing EU consumption, the EU has increased supplies on world markets and depressed prices. This has clearly been disadvantageous to major exporters such as the United States, Australia, and New Zealand. Superficially, it would appear to have been beneficial for food-deficit Third World countries dependent upon imported cereals to feed their populations. However, many development economists would argue that the long-run microeconomic impact of the disincentive effect of low world market prices on the governments of many developing countries to liberalize their own distorted food markets has outweighed the short-run macroeconomic benefits of a reduced import bill.

As EU production has expanded in the face of static consumption, traditional exporters have first seen their share of the EU market contract, subsequently found their products expelled from the EU market, and then seen subsidized EU exports compete away their market share in other countries. Thus, although the initial intent of the Uruguay Round was simply to reduce agricultural protection overall, in the face of EU intransigence its trading partners switched to a strategy of trying to impose GATT disciplines on the CAP's variable import-levy and export-subsidy mechanisms.

World market price instability stems from the EU's pursuit of domestic price stability—arguably one of the objectives of Article 39(1) EEC. Variable import levies and export subsidies certainly act to insulate EU producers and consumers from world market price movements: that is what they were designed to do. To that extent, any supply shortfall or demand surge experienced in the world economy will have to be absorbed by producers and consumers other than those located in the EU. Thus, the price movements necessary to bring about the requisite adjustments will be greater than would have been the case had the EU's citizens also been subject to the price change.[8]

Set against this is the fact that the EU periodically accumulates major stockpiles of intervention goods, which if managed in a countercyclical fashion could offset the otherwise price-destabilizing effects of the CAP. However, there is no evidence to suggest that the EU does consciously manage its stocks in this way.

CAP Decision Making

The EU has five institutions: a parliament, a council, a commission, a court of justice, and a court of auditors. As we noted in the preface, the Council now styles itself as the Council of the European Union, and the Commission as the European Commission. For many years the heads of state or government have met every six months to agree between themselves upon new policy initiatives and to resolve conflicts. These summit meetings are known as meetings of the European Council, and although the European Council has no legislative powers under the EC treaties, it has played an important part in the development of the EU's policies, including the CAP.

All EU legislation relating to the CAP, though not necessarily all EU legislation of relevance to the farm and food industries, is derived from Article 43(2) EEC. This states that

> The Council shall, on a proposal from the Commission and after consulting the European Parliament, . . . by a qualified majority . . . make regulations, issue directives, or take decisions, without prejudice to any recommendations it may also make.

This classic formula, which for the CAP has remained unchanged since 1958, gives the Council and Commission significant powers. The Commission *proposes,* and the Council *disposes.* All policy initiatives come, in theory, from the Commission; and without a formal Commission proposal the Council is powerless to act. Council decisions are to be

---

8. See, for example, Johnson (1975) for an elaboration of this theme.

reached by qualified majority vote, whereas in some more sensitive areas of EC business unanimity is the rule. The formal role of the European Parliament in adopting CAP legislation is limited to an advisory capacity, although its powers over the budget are quite significant, and in other areas of legislation—for example, the harmonization of food law—it now has an enhanced role to play. Finally, it should be noted that over the years the European Court has assumed a central role in shaping many areas of policy as, in applying EU law, it has sought to interpret the treaties and derived legislation.

The Council

Although its powers are limited by the treaties, the Council is the center of the EU's decision-making process. It consists of representatives from each of the member states meeting to discuss matters of relevance to the treaties. In practice, the agendas for meetings of the Council are sorted by subject. CAP matters for discussion will appear at a particular meeting of the Council, to which the member states will send their ministers of agriculture. Hence, particular meetings of the Council will be known as the Agriculture Council, the Environment Council, and so on.[9]

Chairmanship (the presidency) of the Council rotates on a six-month basis, with its members shifting clockwise around the Council table. While the minister takes the chair, his or her deputy continues to represent the member state. Each presidency listed in table 3.3 starts on 1 January or 1 July, and it imposes a significant administrative burden on the civil service of the member state concerned, even though the Council is served by its own sizable secretariat. Some member states, it is felt, are better able to cope with this administrative burden than others.

Until December 1994 there was a twelve-year cycle in operation: during the first six years of this cycle the member states took the presidency in alphabetical order. Thus, it was the Portuguese agriculture minister, Arlindo Cunha, who chaired the meetings on the CAP reform package in May 1992. In the second six years, the order was reversed in each twelve-month period, reflecting the fact that less work tends to be done in the "holiday" months of July, August, and September.

There are, of course, good and bad chairmen; and in part the success of a particular presidency will rest upon the personal abilities of that member state's ministers. But the presidency can also take a more active

---

9. The terms *Agricultural Council, Farm Council, Council of Agricultural Ministers,* and so on, are used interchangeably.

role in influencing EU business during its six months of office. For example, with the Council secretariat, it can determine the agenda of Council meetings. Thus, if a particular matter is not urgent, it can be pushed by an enthusiastic presidency or ignored by an unenthusiastic

TABLE 3.3.   Presidency of the Council of Ministers[a]

1982	I	Belgium
	II	Denmark
1983	I	Germany
	II	Greece
1984	I	France
	II	Ireland
1985	I	Italy
	II	Luxembourg
1986	I	The Netherlands
	II	United Kingdom
1987	I	Belgium
	II	Denmark
1988	I	Germany
	II	Greece
1989	I	Spain
	II	France
1990	I	Ireland
	II	Italy
1991	I	Luxembourg
	II	The Netherlands
1992	I	Portugal
	II	United Kingdom
1993	I	Denmark
	II	Belgium
1994	I	Greece
	II	Germany
1995	I	France
	II	Spain
1996	I	Italy
	II	Ireland
1997	I	The Netherlands
	II	Luxembourg
1998	I	United Kingdom
	II	Portugal

Source: Article 2 of the Treaty establishing a Single Council and a Single Commission of the European Communities, as amended.

[a] This cycle was amended as a result of the accession of Austria, Finland, and Sweden on 1 January 1995.

one. Most CAP price support matters, however, are so pressing that they cannot be ignored.

The conduct of a Council meeting can also influence its outcome. With each minister, each wanting his or her say, Council meetings can be extremely lengthy. Under such circumstances, it is easy enough, for a reluctant presidency, to drag discussion out from one Council meeting to another, whereas a presidency determined to reach a decision could keep the meeting in continuous session with all-night sittings several days at a stretch.

It is often suggested that the Council of Agricultural Ministers acts as a Council *for* Agriculture, possibly at variance with the wishes of the member states. While it must be recognized that each minister in the Council is beholden to his or her cabinet colleagues back home, the Agriculture Council does have some of the characteristics of a club. It meets regularly, more frequently in fact than most other councils, and often for late, all-night sittings. Furthermore, its members tend to have long tenures. The former German minister Ignaz Kiechle, for example, held office continuously from March 1983 to January 1993, and his predecessor had served since 1969. The former Portuguese minister Arlindo Cunha had been attending the Agriculture Council since 1986, first as secretary of state and then as minister. A short-serving minister of culture might never sit in a Council; a minister of agriculture, however, will probably know his or her fellow Council members well.

Also present at a Council meeting will be a Commission representative, usually the appropriate commissioner. It is the commissioner's responsibility to present and explain the Commission's proposal to the Council and if necessary—after consulting his or her fellow commissioners—to amend the proposals. Often, during long and detailed discussions, the presidency will present a compromise proposal to the Council. Whether or not the Commission is willing to endorse, and in effect adopt as its own, the presidency's compromise can critically affect the outcome of the discussion.

The Council cannot legislate without first receiving a proposal from the Commission: in the absence of a proposal, it is powerless to act. It can accept a Commission proposal on the basis of a qualified majority vote, which we describe in the next section. However, if it wishes to amend the Commission's proposal a unanimous decision is required.[10] Thus, the presidency and the Commission will work closely together

---

10. Abstentions, or absences from the Council chamber, do not thwart unanimity. A member state would have to vote against the proposal for it to be blocked.

during complex negotiations because if the presidency can identify a compromise proposal acceptable to the Commission, it can then be adopted by qualified majority vote.

Qualified Majority Voting

The provisions for qualified majority voting pervade the treaties and are not unique to the CAP. The weights that member states' votes had until 31 December 1994 are as follows:

Belgium	5
Denmark	3
Germany	10
Greece	5
Spain	8
France	10
Ireland	3
Italy	10
Luxembourg	2
Netherlands	5
Portugal	5
United Kingdom	10

Thus, the total is seventy-six, and a qualified majority amounts to fifty-six. If member states with a weighting of twenty-three or more fail to cast their votes for a proposal, for whatever reason, this would constitute a blocking minority, and the proposal would fail.[11]

The voting strengths are a political compromise and only very roughly related to population size and economic importance. Luxembourg is generously treated with a weighting of two. Germany, following its reunification, promised not to ask for an increase in its weighting for fear of unbalancing the political compromise that initially established

---

11. The Council's own procedural rules, however, specify that a qualified majority vote can only take place if a proposal has been on the agenda for fourteen days, unless the Council unanimously decides otherwise. This became an issue in June 1992 during the adoption of the legal texts implementing the Mac Sharry reforms. In the event, in the early hours of 1 July 1992, with the clock stopped at midnight on 30 June, the Portuguese presidency secured unanimous agreement on the package of regulations. Once 1 July 1992 had officially arrived, under a British presidency and with the fourteen-days' notice met, the package could have been adopted by qualified majority vote—but Portugal would have lost face had this happened.

the voting strengths. The qualified majority has been pitched at a level that does not allow the four (or five if Spain is included) large member states to push through a measure without the support of some of the smaller member states.

The accession of Austria, Finland, and Sweden on 1 January 1995 extended the total number of votes to eighty-seven—with three votes for Finland and four each for Austria and Sweden. In March 1993, following an acrimonious debate in which the United Kingdom tried to maintain the blocking minority at twenty-three following this proposed enlargement, the Council decided to increase the blocking minority to twenty-six in the European Union of fifteen states. An intergovernmental conference to be convened in 1996 will return to the vexed issue of EU decision making.

A blocking minority could arise under four circumstances:

• a vote against the proposal;
• an abstention;
• lack of representation in the Council (though a Council member can act on behalf of another member state); or
• a refusal to take part in the vote.

Historically, the last circumstance—refusal to take part in the vote—held a special significance, and it meant that member states willing and able to invoke the "Luxembourg Compromise" had power of veto. The Luxembourg Compromise dates back to the mid-1960s. Basically it is, or was, an informal compact between certain member states to support each other if one of the group declared that the matter under discussion in the Council was of such significance to the country concerned that it could not afford to be outvoted. France, Denmark, the United Kingdom, and Greece certainly took this position for a good many years. If one of these four asked its colleagues not to take part in a vote under the Luxembourg Compromise, then a blocking minority was secured, and the measure could not pass. Thus, for those countries willing to invoke this measure, the Luxembourg Compromise gave it power of veto in the Council, notwithstanding the treaty provisions for qualified majority voting.

As a consequence, throughout the 1970s and into the early 1980s, EC decision making, except in the Budget Council, was not subject to majority voting. Consensus decisions were the norm. In the case of CAP policy making this meant the careful construction of elaborate package deals in which the whole package stood or fell on the basis of a mutual acceptance of every constituent part. A member state with a particular vested interest could hold the rest for ransom, until its interest groups had been bought off. Radical decision making was not possible, and the CAP ran out of control.

Since 1982, qualified majority voting has increasingly displaced consensus decision making and is now the norm in the Agriculture Council—though it must be conceded that in the 1980s there was no marked improvement in the quality of CAP policy making. The Luxembourg Compromise was last used in the Agriculture Council by Greece in 1988. Whether or not a member state could now successfully invoke the Compromise is a delicate, and unanswered, question. As we shall see later in this text, in the closing phases of the Uruguay Round its use was certainly threatened.

The Commission

Although in some respects the Commission is the EU's civil service, it is both more and less than this. It certainly has a strong political role, as evidenced by the high public profile of Commission President Jacques Delors. It is not quite the civil service of Europe in the sense that much of the detailed implementation of EU policy is undertaken by nationally employed, and paid, civil servants: the CAP is implemented in Britain by the Ministry of Agriculture, Fisheries and Food; the Intervention Board; and Her Majesty's Customs and Excise, for example.

Its political role stems from the fact that all new policy initiatives, within the framework of the treaties, are supposed to originate from the Commission. As we noted previously, in the absence of a Commission proposal the Council is powerless to act. Similarly, if the Commission withdraws a proposal previously submitted to the Council, then the Council is unable to legislate.

Furthermore, the Commission is the guardian of the treaties and of all the legislation enacted under the treaties. If the Commission believes that a member state is failing to implement some piece of EU legislation, then it is the Commission's responsibility to take action against that member state, ultimately in the European Court.

The word *Commission* can take on two meanings. It can be used to refer to the thirteen thousand-plus permanent employees who work in the Commission Services or to the seventeen commissioners who head the Commission.[12] The commissioners have to be nationals of the member states, and there must be at least one (and no more than two)

---

12. Twenty from 1 January 1995.

national from each member state. Commissioners should be "chosen on the grounds of their general competence," and their independence is to be "beyond doubt." Moreover, in undertaking their duties, "they shall neither seek nor take instructions from any Government or from any other body."

In practice, the names come from the member states: two from the five large member states and one from each of the others. Although the choice of the Commission president can be the subject of considerable debate and individual nominations for the presidency vetoed, member states in effect choose "their" commissioners. Thus, if a commissioner wants to secure a second (or more) term of office he or she might be well advised to placate the home government regardless of the theory that commissioners take instructions from no one. Equally, a desire to return to a political career in one's home country after serving as a commissioner might influence current thinking and actions.

At the beginning of their period of office, the new commissioners generally meet in an isolated retreat where they can allocate responsibilities between themselves. The Commission services are organized into a number of directorates-general (known as DGs) in rather the same way that a national civil service would be arranged into ministries. In the 1993–94 Commission headed by Jacques Delors,[13] René Steichen of Luxembourg had responsibility for agriculture and rural development and as such was in charge of the Commission's directorate-general for agriculture. Sir Leon Brittan was commissioner for the EC's external trade relations and as such had overall responsibility for the GATT talks. The commissioners act as a collegiate body and at their weekly meetings take decisions by simple majority vote. Each commissioner appoints a small office staff, or cabinet, on which he or she will rely for advice and expertise on matters not covered by the directorates-general for which he or she has responsibility. Because the expertise of the directorate-general for agriculture is captured by the commissioner for agriculture, it is difficult for the other commissioners to have much control over the CAP if their cabinets are unable to provide adequate advice.

Council and Commission

Although the Parliament, the Court of Justice, and the Court of Auditors have a part to play in the formulation and implementation of the

---

13. As a consequence of the changes introduced by the Maastricht Treaty, the 1993–94 college of commissioners served for only two years. As of 1 January 1995, the college of commissioners will serve a five-year term.

CAP, the key players are the Council and the Commission. As noted earlier, formally at least all policy proposals must be made by the Commission—though the Commission will, of course, often be very receptive to ideas developed in the Council and particularly those floated by the presidency. New policy initiatives, if accepted by the Council on qualified majority vote, will take the form of Council regulations that have direct force of law throughout the EC once they are published in the *Official Journal of the European Communities.* Usually, however, day-to-day running of the policy will be delegated to the Commission; and it will be the Commission that determines import levies, sets export refunds, and administers the intervention arrangements. In delegating many of these responsibilities to the Commission, however, the Council has insisted that the Commission may only act after consulting the management committee for the product in question.[14] The Council and Commission have been known to bicker over the boundaries lying between their respective spheres of competence.

The Common Commercial Policy and Commission Competence

The EEC Treaty established a common market with a common commercial policy, which was to be based "on uniform principles, particularly in regard to changes in tariff rates, the conclusion of tariff and trade agreements, the achievement of uniformity in measures of liberalisation, export policy and measures to protect trade such as those to be taken in case of dumping or subsidies" (Article 113(1) EEC). Article 113(3) set out the procedures for negotiations: the Commission made a recommendation to the Council, the Council on a qualified majority authorized the Commission to open negotiations, the Commission conducted the negotiations "in consultation with a special committee appointed by the Council to assist the Commission in this task and within the framework of such directives as the Council may issue to it," and then Article 114 EEC provided for the Council to conclude such agreements, acting on a qualified majority. Thus, a permanent Article 113 committee became one of the de facto institutions of the EU. Article 229 EEC gave the Commission responsibility for ensuring "all appropriate relations with . . . the General Agreement on Tariffs and Trade."

However, it was the twelve member states that were members of the 1947 GATT, and although the Commission led the negotiations in

---

14. For further discussion of CAP decision making, see Swinbank (1989).

the Uruguay Round the separation of powers between the Commission, the Council, and the member states in GATT negotiations has never been entirely problem-free. The Commission would like to think it has sole authority to negotiate on behalf of the EC, while the Council would claim it has only limited authority to do so. And the member states, pointing out that they rather than the EC were GATT members, would claim residual rights to participate in GATT deliberations, notwithstanding Article 113 EEC.

The Maastricht Treaty which came into force on 1 November 1993 resulted in some changes. Article 114 EEC has been repealed, and instead Article 228 EC deals with the procedures relating to the conclusion of trade agreements. One possible interpretation of the new Article 228 EC is that the procedures are unchanged in that Article 113(3) EC agreements are particularly identified. However, "agreements having important budgetary implications for the Community" can now only be concluded "after the assent of the European Parliament has been obtained," and it was difficult to argue that the Uruguay Round agreement would not have important budgetary implications, not just for the CAP but also with respect to reduced tariff revenues following the lowering of trade barriers. In the event, the European Parliament staked its claim to be consulted on the conclusion of the Uruguay Round agreement, and the Council subsequently conceded it would seek the parliament's assent (Council of the European Union 1994, 18). Also problematic was the fact that the Uruguay Round agreement included services and intellectual property rights. Some argued that these matters went beyond the EU's competence, which relates solely to trade issues, and remain the sole prerogative of the member states.

The Power of the Farm Lobby

In 1958 the citizens of the six member states of the fledgling EEC remembered only too well the postwar privations of food shortages, and many townsfolk had relations on the land. In all member states except Belgium, over 10 percent of the workforce was engaged in agriculture (see table 3.4), and the voting strength of the farm sector was strong. It is perhaps not surprising then that the EC began constructing a CAP with strong protectionist tendencies, drawing upon the policies already in place in the member states.

Nearly forty years later, the power of the EU's farm lobby is not as easily explained. There has been a massive exodus from the land, and as table 3.4 shows the relative share of agriculture in overall employment is much diminished. Simply in terms of numbers of votes, it cannot be said that the farm lobby retains its previous importance. In part, however, table 3.4 overstates the case because jobs have also been shed from the farm sector, and some tasks previously undertaken on farms are now performed by the input and first-stage food processing industries that have a vested interest in maintaining the CAP.

In Ireland, Greece, Portugal, and Spain, the farm vote remains considerable. In France, it is regionally important, and the French farm lobby has been permitted by the state to protest vociferously and violently to further its claims. The nature of coalition politics in West Germany meant that for three decades the agriculture ministry was in the hands of the minority coalition party with an agricultural constituency. In Denmark, too, the farm vote has retained an importance. And Denmark shares with Ireland the fact that as major exporters of agricultural products, their economies receive a considerable financial boost by budget and other transfers generated by the CAP.

Another factor that works in favor of the status quo is the nature of decision making under the CAP, outlined earlier in this chapter. Policies tend to be enacted forever: they run until amended, and reform can be difficult when decisions have to be unanimous (as was in practice the case until 1982) or if a small blocking minority can thwart change. The evolution of the CAP might have been very different if the policy measures had only been enacted for five-year terms, after which they lapsed unless renewed. The need to put together a qualified majority in favor

TABLE 3.4.   **Percentage Share of Agriculture in Total Civilian Employment**

	1958	1970	1980	1990
Belgium	9.4	5.0	3.2	2.7
West Germany	15.7	8.6	5.3	3.4
France	23.7	13.5	8.7	6.0
Italy	34.9	20.1	14.2	8.8
Luxembourg	17.9	8.8	4.8	3.3
The Netherlands	12.6	6.2	4.9	4.6
Denmark	15.9	12.9	8.0	5.5
Ireland	38.4	27.3	18.4	15.0
United Kingdom	4.4	3.2	2.6	2.1
Greece		40.8	30.3	23.9
Spain		27.1	19.3	11.8
Portugal		30.0	27.3	18.0
EC–12		13.5	9.6	6.5

Source: Eurostat (1973, table A2) and Commission (1993a, table 3.5.1.3).
Note: Includes self-employed farmers and unpaid family workers.

of a renewal of a measure would have strengthened the hand of the reformist lobby.

Furthermore, the policy is incredibly complex and difficult to explain to the public (and, one suspects, to politicians). The "managed market" mechanisms, under which grossly distorted farm prices deliver transfers from consumers to the farm sector, are the very antithesis of a transparent policy mechanism. The public finds it much easier to accept the false claims of the farm lobby that assert that only by maintaining a prosperous farm sector can security of an adequate volume of safe and wholesome food supplies be assured and the environmental features of the countryside maintained.

Surpluses, Budget Overspends, and CAP Reform

From the 1950s onward, EU farm production has grown at a faster rate than consumption, increasing self-sufficiency ratios for virtually every farm product imaginable. Much of the increased output has come from increased yields per hectare, or output per animal, rather than from increases in the cultivated area or number of livestock. The increased yields in turn broadly reflect productivity gains in the farm sector. Nonetheless, it is undoubtedly true that the high levels of price support under the CAP have encouraged production and discouraged consumption and have thus misappropriated the productivity gains.

There is a widespread misconception in EU farming and farm policy-making circles that productivity gains lead inexorably to higher yields, growing surpluses, and a surplus of fertile farmland. This is a misleading conclusion. At one extreme, with unchanged support prices, productivity gains would lead to a larger output using the same resource base. At the other extreme, in the face of a perfectly price-inelastic demand generating a fall in market prices, the same level of output could be produced with a smaller resource base. A competitive market scenario would result in some increase in output that could be sold competitively in the marketplace and some reduction in the resources devoted to farm production. The CAP, in seeking to sustain farm incomes, tends to maintain more people in the farm sector than would otherwise remain, thus generating unwanted surpluses.

The evolving crisis can be illustrated by referring to figure 3.3, which depicts developments in the EU cereals market. Similar graphs can be drawn for all other CAP products. In the early 1970s, the EU was a net importer of cereals, and on the whole the import-levy revenue generated by the policy covered the cost of intervention buying and export refunds. However, production continued to grow while consump-

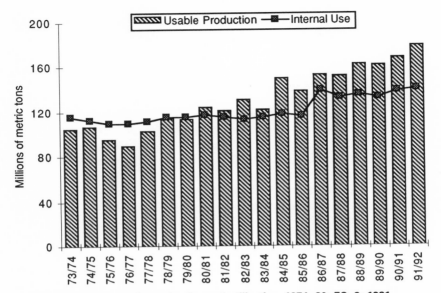

Figure 3.3. EU cereals production and utilization, 1974–80: EC–9; 1981–86: EC–10; 1987–90: EC–12; 1991–92: EC–12 plus East Germany. (Data from Eurostat and the European Commission.)

tion stagnated, and at the end of the 1970s the EU became a net exporter.[15] This has two adverse consequences. First, the budgetary costs of the policy soar because exports have to be subsidized, generating internal political constraints to the policy. Second, the international trading community objects vociferously to loss of market share in the face of subsidized competition from the EU.

In the late 1970s, it was widely believed that the budgetary pressures stemming from market imbalances in the cereal and other commodity markets would force some sort of CAP reform. This was because the budgetary funds available to the EU were strictly limited and little prospect was seen for extending the tax revenues placed at the EU's disposal. However, by a series of accounting and other devices that deferred expenditure to later budget periods and shifted some of the financial burden back to the member states, the EU was able to avoid serious reform of the CAP.[16] Nonetheless, the spendthrift days of the mid-1970s were over,

---

15. Note that in any one year the excess of production over consumption can be added to intervention stores but also that this tends to a be a short-term palliative, with the excess production eventually finding its way to the world market.

16. For details, see Tanner and Swinbank (1987).

and gradually—if reluctantly—limited policy reforms were debated and approved.

In 1980, in its *Reflections on the Common Agricultural Policy,* the Commission advocated a general "principle of co-responsibility" whereby "any production above a certain volume to be fixed . . . must be charged fully or partially to producers" (Commission 1980, 18), and in the 1982–83 marketing year "guarantee thresholds" were first introduced for a number of commodities. In essence, the guarantee threshold mechanism meant that if deliveries or production exceeded the threshold, then automatic price reductions were triggered the following year. Elsewhere we commented, "In the event, the formula for triggering price reductions has not been strictly adhered to and, in our opinion, the ensuing price reductions have proved trivial" (Tanner and Swinbank 1987, 292).

Milk production in 1983 exceeded the guarantee threshold by 6.5 percent, and the Commission reported that "in order to offset fully the additional expenditure likely to arise from the guarantee threshold being exceeded in 1983, the milk price would have to be abated by as much as 12 percent" (Commission 1983, 17). This was judged to be politically unacceptable, and so instead the Commission proposed the introduction of a system of milk delivery quotas. These were enacted by the Council in March 1984.

For cereals, a 1982–83 production threshold of 119.5 million metric tons was set for all cereals except durum wheat. If the average production over the last three marketing years exceeded this threshold, then the intervention price was to be reduced by 1 percent for every million metric-ton excess, subject to a maximum 5 percent cut. A record harvest in 1984–85 resulted in the threshold being exceeded by over 8 million metric tons. It was difficult for the Commission to do other than propose the 5 percent price cut that the guarantee threshold mechanism provided for. However, this was reduced to a 3.6 percent price cut by deducting the "normal" price increase that would otherwise have been proposed! A price cut of 3.6 percent was unacceptable to the German minister, as was a compromise proposal of 1.8 percent. Indeed, Germany successfully invoked the Luxembourg Compromise and blocked a Council decision on this proposal. The Commission, nonetheless, acting under its powers of market management, took the unprecedented step of implementing the 1.8 percent cut in support prices in the absence of a Council decision.[17] These developments did not inspire confidence in the efficacy

---

17. For further details, see Swinbank (1989).

of guarantee thresholds or in the willingness of the Council of Farm Ministers to countenance serious CAP reform. For cereals, guarantee thresholds were abandoned in 1986 and replaced by a coresponsibility levy (a producer tax) initially set at 3 percent of the intervention price. In 1987, we wrote:

> At this stage, the Council does not appear to have the political will to implement market-oriented reform and instead has demonstrated a preference for introducing further regulations which "contain" expenditure (in the case of quotas) or increase the funds available (in the case of the co-responsibility levy). As long as the Council's *ad hoc* policies are sufficient to avoid the consequences of the budgetary constraint, we would argue that the CAP is unlikely to undergo comprehensive reform. Changes to the CAP as a result of changing public attitudes to the excesses of agricultural support are likely to occur only in the longer term. (Tanner and Swinbank 1987, 294)

That remained true for the remainder of the decade, but by then the Uruguay Round of GATT negotiations was underway.

CHAPTER 4

The Negotiations I: From Punta del Este to Heysel

In retrospect it can probably be claimed that the origins of the Uruguay Round lie in a GATT ministerial meeting in Geneva in November 1982. At that meeting, we are told, the United States had hoped that "a decision would be taken . . . to renegotiate a number of GATT rules, including the code on subsidies" (Commission 1984, 60). What was agreed upon was a work program that "set the stage for the launching of a new round of negotiations as and when the political and economic climate made such a decision by governments possible" (*Trade Policies for a Better Future* 1987, 137). In particular, the ministerial meeting decided to establish a Committee on Trade in Agriculture "to examine international trade in agricultural products in the light of GATT rules and the effects of different national policies." At the time, the EC Commission claimed that "The Community is taking an active part in the Committee's work in the hope that it can bring about a return to conditions in which GATT rules can operate smoothly" (Commission 1984, 56). The Commission had expressed considerable aggravation with the stance adopted by the United States, claiming that "Operation of these rules has been considerably upset by the spate of complaints from the United States about fundamental elements of the common agricultural policy" based on "new readings of GATT rules in conflict with the traditional interpretations" (Commission 1984, 56). Throughout 1983 there was a good deal of talk about the risk of a trade war.

Earlier, in May 1982, the OECD's Ministerial Council had asked its secretariat "to analyse the approaches and methods for a balanced and gradual reduction of protection to agriculture in order to help governments formulate appropriate policies within the open multilateral trading system" (Viatte 1990, 294). The report based upon this work was "endorsed and approved" by the OECD's Ministerial Council in May 1987 and subsequently published (OECD 1987, 3). While the OECD's involvement certainly emphasized the need to reform agricultural support systems that distorted international trade and enhanced the importance of the OECD Agricultural Secretariat itself, it did in the early years lead to some confusion. Because a major appraisal of the impact of agricultural policies was being undertaken by the OECD (rather than the GATT) secretariat, this gave rise to the suspicion that the agricultural negotiations might be undertaken in an OECD setting. This was not to be. But the OECD's methodology—namely, the producer and consumer subsidy equivalents (PSEs and CSEs, referred to in chapter 2) did feed through into the GATT negotiations.

The United States, Australia, and Japan were keen advocates of a new round of multilateral trade negotiations, but other countries were less enamored of the idea. For the U.S. administration, an advantage of embarking upon such a round was that the protectionist tendencies of the Congress could more easily be restrained if negotiations were in train. And for Japan, multilateral discussions would ease the bilateral pressures that the United States and EC were bringing to bear in pursuit of amendments to Japanese trade policy (Winters 1990, 1297). It was not however until mid-1985 that a consensus began to emerge that a new round should be launched, although there was little agreement as to the scope or form of the negotiations. At the time, the EC was determined that the mechanisms of the CAP would not be called into question, and this remained the EC's official position throughout the preparatory stages. On 19 March 1985, in a declaration concerning the possibility of a multilateral round of trade negotiations,[1] the Council had declared inter alia that:

As regards negotiations on agriculture, the Community confirms its readiness to work towards improvements within the existing framework of the rules and disciplines in GATT covering all measures affecting trade in agricultural products, both as to imports and as to exports, taking full account of the specific characteristics and problems in agriculture. In the Community's view, agreement on such improvements should be sought within the body specially established to that effect, that is the Committee on Trade in Agriculture.

The Community is determined that the fundamental objectives and mechanisms, both internal and external, of the common agricultural policy shall not be placed in question. (EC Statement to GATT, July 1995, quoted in Commission 1986, 146. See also *Bulletin of the European Communities,* 3–1985, 58).

---

1. For a contemporary view of the *nonfarm* issues, see Cresson (1985). Edith Cresson was then minister for foreign affairs for the French government.

After further preparatory discussion, the forty-first session of the contracting parties meeting in Geneva in November 1985 decided to establish a Preparatory Committee to determine "the objectives, subject matter and modalities" for a new round of multilateral negotiations to be launched in September 1986 (*Bulletin of the European Communities,* 11–1985, 97). In June 1986 the EC's Council of Ministers approved an overall negotiating strategy for the Commission in the forthcoming negotiations, while stating that its declaration of 19 March 1985 remained "in its entirety the basis of the Community's position on the new round of multilateral trade negotiations" (*Bulletin of the European Communities,* 6–1986, 81). Despite the EC's reticence and its commitment to the notion that "the fundamental objectives and mechanisms of the common agricultural policy shall not be placed in question," it was now evident that agriculture was to be included in the Uruguay Round. Without an implicit commitment on the part of the EC to include agriculture, it is doubtful whether the Australian or the U.S. delegations would have been willing to embark for Punta del Este.

Punta del Este

When, in September 1986, the GATT ministerial meeting assembled at the seaside resort of Punta del Este in Uruguay for the formal launch of the Uruguay Round, the scope and terms of the negotiations had still to be finalized, despite the Preparatory Committee's work. Three alternative draft declarations were on the table, and some participants had other views. An intense four days of negotiations followed before the declaration formally launching the Uruguay Round was agreed upon (*Trade Policies for a Better Future* 1987, 142). Significantly, the United Kingdom held the presidency of the EC's Council of Ministers in the latter half of 1986 and was eager that the round should go ahead,[2] as was Willy de Clercq, the commissioner for trade policy. But there was little enthusiasm among other member states for a GATT deal that would embrace agriculture, and as a consequence there was some friction in the EC delegation. Unexpectedly, the French minister of agriculture, François Guillaume, arrived in town to bolster the French team, suspicious that the trade minister would prove too flexible when confronted by British resolve to press for endorsement of the GATT talks. It

---

2. The British minister Paul Channon was credited with having "formulated crucial wording on agriculture" (*Financial Times,* 22 September 1986, p. 6). The role of the presidency was outlined in chapter 3.

seemed that, even at this late stage, the French hoped to exclude the CAP from the negotiations and were threatening to invoke the Luxembourg Compromise. The Americans hinted they might leave town. Ultimately, however, Guillaume accepted the key agricultural elements of the redrafted text, and the Uruguay Round was underway.

With regard to agriculture, the Punta del Este Declaration said that

> The Contracting Parties agree that there is an urgent need to bring more discipline and predictability to world agricultural trade by correcting and preventing restrictions and distortions including those related to structural surpluses so as to reduce the uncertainty, imbalances and instability in world agricultural markets.
>
> Negotiations shall aim to achieve greater liberalisation of trade in agriculture and bring all measures affecting import access and export competition under strengthened and more operationally effective GATT rules and disciplines, taking into account the general principles governing the negotiations, by:
>
> (i)   improving market access through, inter alia, the reduction of import barriers;
> (ii) improving the competitive environment by increasing discipline on the use of all direct and indirect subsidies and other measures affecting directly or indirectly agricultural trade, including the phased reduction of their negative effects and dealing with their causes;
> (iii) minimising the adverse effects that sanitary and phytosanitary regulations and barriers can have on trade in agriculture, taking into account the relevant international agreements.
>
> In order to achieve the above objectives, the negotiating group having primary responsibility for agriculture will use the Recommendations adopted by the Contracting Parties at their Fortieth Session, which were developed in accordance with the GATT 1982 Ministerial Work Programme, and take account of the approaches suggested in the work of the Committee on Trade in Agriculture without prejudice to other alternatives that might achieve the objectives of the negotiations. (*Trade Policies for a Better Future* 1987, 150–51)

Two points should be noted about this declaration. The first, of course, was that the EC had agreed that farm support should be discussed. But, second, that all parties agreed that the negotiations would not focus exclusively on the CAP's variable import levies and export refunds. Instead, "all direct and indirect subsidies and other measures affecting directly or indirectly agricultural trade" would be considered (Commission 1987a, 148). For the EC, this meant that the U.S. deficiency payments program would also be a potential subject for negotiation.

Just prior to the Punta del Este ministerial meeting, in August 1986 representatives of a group of fourteen agricultural-exporting nations had congregated at Cairns, in Queensland, Australia. Thus was launched the Cairns Group, which sought to coordinate its members' views and promote a more liberal trading environment for agricultural products.[3] Initially, its targets were the trade-distorting agricultural policies of both the EC and the United States, as these major powers had been engaged in a disruptive export subsidy war for a number of years. But when the United States itself began to advocate the abolition of agricultural subsidies, the Cairns Group sought to act as an "honest broker" between the two. Australia, seen by most observers to be the Group's leader, made a considerable diplomatic investment in the round that included attempts to convince the EC's academic, business, and political communities of the merits of its case. In 1985, for example, the Australian Government's Bureau of Agricultural Economics had published a substantial, well-researched study on the CAP, which concluded that the EC

> could achieve its desired objectives by pursuing alternative policies which would be both less costly to itself and less damaging to the agricultural industries of other countries. It is hoped that this study will stimulate debate within and outside the European Community on ways in which such policies might be implemented. (Bureau of Agricultural Economics 1985, iii)

Indeed, the then secretary of the Australian Department of Primary Industry had noted:

> In recent years the Australian Government has made a substantial investment in comprehensive studies of international agricultural policies. This investment has not been made for altruistic reasons—it has been undertaken for the ultimate benefit of the Australian

---

3. Argentina, Australia, Brazil, Canada, Chile, Colombia, Fiji, Hungary, Indonesia, Malaysia, New Zealand, the Philippines, Thailand, and Uruguay. Fiji was not a member of the GATT. Oxley (1990) has emphasized the important role that he believes the Cairns Group played in the negotiations.

farmer and the community at large. To achieve these benefits, however, it was recognised that the work should be undertaken independently and in a non-partisan and professional manner. The Bureau of Agricultural Economics met these requirements. (Miller 1987, iii)

Other members of the Cairns Group had a less consistent, and less convincing, policy stance. Canada, for example, while wishing to secure more liberal world trading arrangements for cereals to aid its prairie provinces, was less sure about free trade in dairy products where it could face import competition. Warley (1994, 123) has pointed out that the dairying and poultry-producing interests of Quebec could not readily be challenged, without adding fuel to the separatist debate in the Canadian federation.

The Issues

Agricultural trade, of course, was not the only topic to be discussed; nor did it prove to be the only controversial issue. Fifteen negotiating groups were established, four of which were sector-specific, dealing with natural resource-based products, tropical products, textiles and clothing, and agriculture (Greenaway 1991). It is, essentially, in the last of these groups that the CAP was under attack.

The main, but by no means the only, protagonists were the United States and the Cairns Group, on the one hand, who were basically in favor of a freer trading environment for farm products, and the EC and Japan, on the other, who wished to retain as much protection for their farm sectors as they could.

The discussions generated a new language, with new words and acronyms, including:

decoupling
recoupling
production neutral
tariffication
transparency
rebalancing
AMS and SMU
green, amber, and red boxes

In trying to support the income of farmers, many governments have attempted to increase the price at which farmers can sell their products (as in the CAP) or to reduce the cost of important inputs into the farming process (as with fertilizer subsidies, for example). Rational businesses, in trying to gain the most advantage from the government's largess, will attempt to increase production. Thus, such subsidy schemes directly affect the level of production and disrupt world trade. Decoupling describes the intent of devising methods of farm support that have no effect on production or consumption and that would thus be acceptable to the international community. As such they would classified in GATT as falling within the "green box" of permitted measures and could continue to be paid indefinitely. Even in a highly abstract sense, production-neutral support schemes are difficult to define. Politically, they are vast gray areas; and in 1992 a source of contention was whether the EC's Mac Sharry compensation payments (discussed in chapter 5) and the U.S. deficiency payments, should be deemed to be decoupled, falling within the green box. Economic appraisal of these schemes would suggest not, but political expediency dictated otherwise.

Many agricultural-exporting nations had also expressed concern about the likely effect on world market price stability of farm support policies such as the CAP. As we saw in chapter 3, the CAP's variable import levy and export refund insulate domestic producers and consumers from world market price movements. Thus, in securing domestic price stability—much prized by producers and consumers in the EC—greater international price instability is likely to ensue. Recoupling referred to suggestions that policy mechanisms such as the EC's variable import levy should be outlawed—that is, placed in the "red box" in the GATT classification—and a link reestablished between domestic and world market price formation.

Recoupling thus led on to tariffication. The idea here was that all existing protective measures should be converted into a tariff equivalent. As well as achieving the policy objective of recoupling, this would be a transparent policy mechanism that could be subject to an agreed program of partial (or complete) reduction over a specified period. Transparency is an attribute sought by those wishing to reduce agricultural protection because with transparent policy instruments not only can international negotiators identify an objective variable (the tariff) that can be discussed, but—it is said—citizens too can more readily appreciate the costs they bear as a consequence of protecting the farm sector, and this in turn will increase the political pressure on governments to reduce protectionism.

In time, the EC came to accept the principle of tariffication, but as we discuss later, the Uruguay Round endorsed a weakened version of this concept. The EC had linked tariffication with rebalancing.[4] In the Dillon Round, as we noted in chapter 1, the EC entered into GATT bindings on oilseeds, manioc, and other products. The fact that various carbohydrate and vegetable protein products were able to enter the EC at world market prices, combined with a sophisticated animal feeds industry knowledgeable about animal nutrition and with access to computer technology for formulating least-cost rations, led to the phenomenon of cereal substitutes. The CAP's variable import levy mechanisms, export subsidies, and intervention buying had pushed the EC price of cereals well above world market levels. Consequently, with the CAP's distorted pricing, it was commercially feasible to displace highly priced cereals from animal feeds with a variety of imported feed ingredients, including citrus pulp, maize gluten feed, manioc, and soybean meal. Animals have always been fed the by-products of the food industries, but if prevailing world price ratios had applied within the EC, imports on the scale experienced would not have been commercially attractive. One way out of this dilemma was to restrict imports of cereal substitutes into the EC, driving up their EC price and redressing the adverse price ratio in favor of the inclusion of more cereals in animal feed. Hence, the EC's commitment to rebalancing, which was vociferously resisted by the United States.

If the GATT signatories could be encouraged to agree to tariffication, the United States and the Cairns Group wished to negotiate an agreed reduction in protectionism on three interrelated fronts: (1) the newly established tariff would be reduced by a given percentage over an agreed period, (2) domestic support would be reduced, and (3) export subsidies would also be reduced. These would be the policy mechanisms in the so-called amber box. The EC consistently expressed caution over this strategy because it saw all of the CAP's policy mechanisms as an integrated whole: in this view of the world, a reduction in the level of domestic support would automatically lead to a reduction in the levels of border protection and export subsidies.

But if domestic support, falling in the amber box, were to be subject to an agreed reduction over a specified time, then it had to be measured. One possibility was to use the producer subsidy and consumer subsidy equivalents (PSEs and CSEs) that had been calculated by the OECD. In

---

4. A number of studies have examined rebalancing, including one commissioned by the EC (Koester et al. 1987).

practice, the parties chose to refer to an aggregate measurement of support (AMS), although the EC's first preference had been to use the term *support measurement unit* (SMU).

In the early days of the Uruguay Round there was considerable support for adopting a single measure of support that could be negotiated and bound in GATT.[5] However, as the round progressed most participants came to realize that very different trade implications could follow from the same PSE or AMS, quite apart from the measurement problems inherent in these aggregates. When the United States was pressing for the elimination of all trade-distorting subsidies over a specified time horizon, this was of little importance. However, when it was realized that a much more limited outcome was all that could be expected, the problems inherent in the use of a single measure became more acute. Thus, by 1991, Sumner (1992, 240) claimed that "There is now broad agreement on binding commitments for separate disciplines in each of the following three areas: internal support, market access for imports, and export subsidies."[6] Nonetheless, for a long time the EC continued to press for a single commitment on an AMS, arguing that import access and export limitations would follow automatically from a reduction in domestic support.

A PSE, typically, measures the support given per unit of production (in percentage terms or as dollars per metric ton, for example). As such, it is unable to distinguish between a payment made directly to producers, such as a deficiency payment, and a rigged market price. Under the former policy mechanism, production is artificially enhanced, but consumption, broadly speaking, is unchanged. With the latter policy mechanism, consumption is depressed because of the higher market prices, and the effect of the policy on trade is more marked. Similarly, a policy of unrestrained price support to the farm sector, and the same level of price support paid only on a controlled quantity of output limited by quota, would result in the same per-unit PSE but very different trade effects.[7] Countries would want "credit" for the restraint they had shown in controlling production and in particular in limiting exports, and so a per-unit measure of support seemed inappropriate.

In a blunderbuss fashion, the AMS allowed for production con-

---

5. Tangermann, Josling, and Pearson (1987) favored this approach.

6. Deputy Assistant Secretary for Economics, U.S. Department of Agriculture.

7. A PSE is usually measured in nominal terms and thus can be misleading in the livestock sectors if the price of animal feeds is artificially raised as a result of support to cereal growers. For a more detailed critique of PSEs, see Peters (1988), and of the different trade implications of alternative policies, see Harvey (1994).

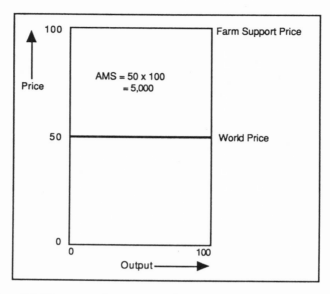

Figure 4.1. Diagrammatic representation of the aggregate measurement of support

trols because an AMS was a measure of the financial support received by a particular farm sector, such as livestock production or agriculture as a whole. Figure 4.1 illustrates the AMS concept diagrammatically. It is essentially the difference between the domestic and the world price, multiplied by the volume of production, plus any direct or indirect payments received by the farm sector. If the world price stands at 50 percent of the domestic price level, then a 10 percent reduction in the AMS could be achieved through a 5 percent reduction in domestic price levels or a 10 percent reduction in production, which could, for example, be brought about by quotas or set-asides. This was an attractive alternative for the EC, but again the foreign trade implications clearly depend upon the particular details of the policy mechanisms in place. As Sumner (1992, 240) pointed out, GATT is about trade rather than domestic policies. The contracting parties are not really concerned about the economic welfare of their trading partners, and so it was perhaps inevitable that the negotiations would eventually focus on import-access and export-subsidy constraints, leaving countries free to maintain domestic policy mechanisms that continue to impose considerable costs upon domestic taxpayers and consumers and the economy in general.

The Zero Option: From Montreal to Brussels

The Uruguay Round was set to last for four years, with a mid-term review in Montreal in December 1988 and a ceremonial conclusion in Brussels in December 1990. As far as agriculture was concerned, this was an ill-fated timetable. A final deadline for the round was necessary to meet the peculiarities of the United States' legislative procedures. Under the so-called fast-track procedure the U.S. president had until 3 March 1991 to sign an agreement, which would then have to be ratified by Congress. If he met this deadline, Congress was committed to accepting or rejecting the package as a whole. If he missed the deadline, Congress would have the right to examine the agreement item by item, thus potentially unscrambling the package (Montagnon 1990).

The United States' opening bid was for what came to be known as the "zero option." This would have involved the phased elimination, over a ten-year period expiring in the year 2000, of all trade-distorting farm policy mechanisms (Josling 1991, 272; Hine et al. 1989, 386).

By contrast, no longer under the influence of a British presidency, the EC's offer was very limited in scope and indeed would have extended onto a world scene the "managed market" concept of the CAP. Furthermore, the communiqué issued after the EC Council meeting of 19–20 October 1987, which had announced the agreement on the EC's negotiating strategy, had stressed once again that "the fundamental mechanisms of the CAP must be preserved" (*Agra Europe,* 30 October 1987, E2).

There has been considerable speculation as to the Americans' motives: the United States had not previously been noted for its free trade philosophy in the agricultural sector. Paarlberg (1991, as noted in Hillman [1994], 32) has suggested that the zero option was in fact an attempt by the Reagan administration to pursue a domestic policy of liberalization, forcing deregulation on U.S. farmers after Congress had rejected the administration's free market proposals in the 1985 Farm Bill. Paarlberg further suggests that "certain protectionist interests in the United States nevertheless supported the 'zero-option' in the secure knowledge that it would be rejected by the other parties to the negotiations and that it might possibly torpedo the entire Round" (Hillman 1994, 33–34).

In retrospect, it would seem that the U.S. view of decoupled payments would have ensured that their deficiency payments could have continued under the zero option. But politically the damage was done. The Europeans could not take the American proposal seriously; it had to be an outrageous bluff. The United States, for its part, failed to appreciate the limited room for maneuver that the EC policy-making framework allowed and continued to do so through December 1990. It is quite remarkable that these two partners could have been so badly misinformed about each other's intentions.

With a lame-duck presidency and a lame-duck Commission, little more than impasse could be expected of the Montreal mid-term review.[8] Nonetheless,

> tentative agreement was reached in eleven of the fifteen negotiating groups . . . , but not in four of the groups. These covered agriculture, intellectual property, textiles and clothing, and safeguards.
>
> In the agricultural group, the marked divergence in the EC and US negotiating positions with respect to the phasing out of agricultural support led to a stalemate, largely because the parties could not agree on the wording for the ultimate goal of the agricultural negotiations. As a result, the Latin American members of the Cairns Group threatened to refuse ratification of the agreements already reached in the above-mentioned eleven negotiating groups—a course of action which could have brought the entire Uruguay Round to a complete halt. (Riethmuller et al. 1990, 7)

Deft diplomacy on the part of GATT Director-General Arthur Dunkel brought the Negotiating Group on Agriculture to Geneva in April 1989 and resuscitated the talks. The parties agreed that

- "the long-term objective . . . is to establish a fair and market-oriented agricultural trading system";
- the long-term objective was for "substantial and progressive reductions in agricultural support and protection sustained over an agreed period of time," as measured by an aggregate measurement of support;
- "credit would be given for measures implemented since the Punta del Este Declaration which contribute positively to the reform programme";
- "all measures affecting directly or indirectly import access and export competition" would be included;

---

8. President Bush was to take over from President Reagan in January 1989. A new college of commissioners was to take office in January 1989 and, in particular, Frans Andriessen would cease being commissioner for agriculture and instead take on responsibility for external relations.

- participants would make detailed proposals along these lines by December 1989; and
- support levels, expressed in national currencies (or ecu in the case of the EC), would not be raised "above the level prevailing at the date of this decision."[9]

Both the EC and the United States had shifted their positions. The EC had agreed to "substantial and progressive reductions in support," but from the EC's perspective it was a favorable outcome. Because participants would be given "credit" for any measures implemented since the Punta del Este Declaration, significant reductions in the AMS could already be demonstrated, given the high 1986 AMS on which the Commission would subsequently base its calculations. As a result of the support "standstill" being expressed in ecu, some member states could still increase their support levels expressed in national currencies through the simple device of devaluing their green conversion rates. Furthermore, the fact that CAP prices were expressed in green ecu rather than real ecu meant that a certain element of fudge could confuse the issue. The United States, it would appear, had accepted that something less than the zero option would be the outcome of the negotiations. Nonetheless, in its Comprehensive Proposal of October 1989 the United States stuck to the zero option (Hillman 1994, 40).

In its December 1989 submission, the EC failed to specify the extent to which it was willing to reduce price support under the CAP. But it did declare the following:

As foreseen by the Decision in Geneva in April 1989, reductions would be measured against the reference of 1986, in order to give credit for the measures which have been adopted since the Declaration at Punta del Este. (GATT 1989, 4)

It expressed "fundamental" reservations about tariffication but went on to say,

However, the Community is prepared to consider including elements of tariffication in the rules of external protection given that the problem of rebalancing can be solved in the context of tariffication. This could be envisaged on the following basis:

---

9. The April 1989 mid-term agreement is reproduced in appendix A of Riethmuller et al. (1990). The original source cited is *GATT Focus Newsletter,* no. 61, May 1989.

- border protection for the products included on the list of Support Measurement Units, as well as their derivatives and substitutes, would be assured by a fixed component. This component, expressed as an absolute value, would be reduced at a similar rate as the Support Measurement Unit. It would be completed by a corrective factor in order to take into account exchange rate variations and world market fluctuations which went beyond certain limits to be agreed;
- deficiency payments would be treated in the same way and converted into tariffs;
- the same arrangement would apply to exports, the amount granted to exports could not exceed that levied on imports. (GATT 1989, 6)

Although the GATT negotiators had tabled their proposals by December 1989, as 1990 wore on agreement proved as elusive as ever. A 15 October 1990 deadline was set for tabling final offers, which the EC's decision-making bodies were ignominiously unable to honor. Commissioner Mac Sharry's proposal was a far cry from the original zero option of the United States. It envisaged a 30 percent reduction in the AMS over a ten-year period, backdated to 1986.[10] At the time *Agra Europe* (28 September 1990, P1) pointed out that this would

> minimise any actual price cuts which it will have to make in the 1991–96 period. . . . Commission calculations indicate that the production reduction effect of the dairy quotas is worth 20 percent to 25 percent. This means therefore that dairy support would only have to be reduced by 5 percent to 10 percent in the 1991–96 period.

Judging from press reports, the Commissioners' meeting of Wednesday, 19 September 1990, at which Ray Mac Sharry's plans were to be endorsed, was stormy. The draft was rejected: *Agra Europe* (21 September 1990, E1) reported that the commissioners had not considered it "radical enough, particularly in the sensitive area of export refunds" and that Frans Andriessen had led the opposition. This meant there would

---

10. The figure of 30 percent had first been mentioned by Commissioner Mac Sharry at an informal meeting with the agriculture ministers of Australia, Canada, Japan, and the United States in the summer in Ireland (*Agra Europe,* 3 August 1990, E1).

be no GATT offer for the Agriculture Council to discuss the following week, placing the GATT timetable in jeopardy (E1). The deputy president of Britain's National Farmers' Union apparently said he "was 'shocked' that the Commission did not consider that a 30 percent cut in farm support went far enough. 'The idea that the EC should make further concessions on export restitutions is simply not on,' he said" (E2). Meanwhile, from the United States came reports that the U.S. government was abandoning its zero option (P2).

Mac Sharry's revised proposal was not discussed at the commissioners' meeting of 26 September 1990, but it was endorsed "unanimously" late on Wednesday, 3 October 1990 (*Financial Times,* 5 October 1990, p. 28). Some of the details had changed. But the key element of "30 percent over ten years, backdated to 1986" survived. By now the U.S. offer was running late. According to *Agra Europe* (5 October 1990, P1) it was "still undergoing extensive massaging in the Washington political machine."

An additional Agriculture Council had been convened for Monday, 8 October 1990, in Luxembourg, to discuss the Commission's draft. It ended in deadlock, with a "majority of ministers" arguing that "the plan for a 30 percent reduction in farm subsidies could only be acceptable if at the same time the Commission provided a firm promise of structural aid to help the producers hardest hit by the market support cuts" (*Agra Europe,* 12 October 1990, E4). Here was an early pointer to the Mac Sharry reforms.

Students of EC politics should note that EC trade ministers met in council later that week. They were qualified to reach decisions on the EC's negotiating mandate, but they had clearly ceded responsibility to their farm ministerial colleagues. *Agra Europe* (12 October 1990, E4) reported that the Italian farm minister had said that "it had been agreed between himself and the Italian trade minister that the issue would remain under discussion by the agriculture ministers."

Farm ministers reassembled on Monday, 15 October 1990, but a "thirteen-hour session on Tuesday ended in disarray, with ministers unable to agree a formula which would enable them to approve the Commission's offer of a 30 percent cut in farm subsidies over ten years" (*Agra Europe,* 19 October 1990, E1). The United States and the Cairns Group did, however, table their similar offers: the United States called for a 75 percent cut in "the most trade-distorting domestic subsidies" and a 90 percent cut in export subsidies over a ten-year period beginning in 1991 (E4).

The Council meeting was "suspended," not closed, and reconvened on Friday, 19 October 1990. In the meantime,

The official line following Wednesday's cabinet meeting in Bonn was that Germany "must make its contribution" to the Uruguay Round process, and that the GATT negotiations . . . "must not be allowed to fail." But these promises sit uneasily alongside the continuing determination of German farm minister Ignaz Kiechle, who faces a national election in six weeks time, not to sacrifice his farmers on the altar of free trade. (*Agra Europe,* 19 October 1990, P3)

It was reported that the German chancellor had "phoned the Commission President with a warning that the Commission's proposals, as they currently stood, were unacceptable" and that the farm minister would be

seeking from the Commission concrete guarantees that losses in farmers' incomes resulting from the GATT price cuts would be compensated by additional production-neutral aids, and by reinforced set-aside and extensification programmes. (*Agra Europe,* 19 October 1990, P3)

Predictably, the reconvened Farm Council failed on 19 October 1990 to endorse the Commission's plans, as did the foreign ministers on 23 October (*Financial Times,* 20, p. 2, and 24 October 1990, p. 3). For the next attempt, the Commission proposed to compensate farmers "in the form of direct income supports that do not stimulate production" (*Financial Times,* 24 October 1990, p. 3). Agriculture and trade ministers were summoned to a specially convened council in Luxembourg on Friday, 26 October 1990, immediately prior to the weekend meeting in Rome of the European Council. Margaret Thatcher's reaction was characteristically caustic when, after sixteen hours of discussion in Luxembourg, the farm and trade ministers had failed to agree, and Giulio Andreotti—the Italian prime minister—had refused to allow a discussion of the matter in the European Council (*Financial Times,* 29 October 1990, p. 1).

Finally, late on Tuesday, 6 November, the farm and trade ministers did agree on a negotiating mandate. This retained the Commission's key 30 percent proposal but with a number of modifications designed to meet ministers' concerns. For example, the French had been very worried about the impact on EC preference if import protection were reduced at a faster rate than domestic support (*Agra Europe,* 9 November 1990, E1). The EC's offer (discussed in the next section) was tabled on 7 November 1990 (European Community 1990). The evolution of the U.S. and EC positions over the period 1987–90 is summarized in table 4.1.

Four years after the Punta del Este Declaration, only a month remained for serious discussion before ministers assembled at the Heysel Stadium in Brussels on Monday, 4 December 1990, for the formal close of the round. The EC and the United States were almost as far apart in their positions as they had been at the outset. It was clearly an impossible task, and, not surprisingly, it ended in impasse. Matters were not eased by the fact that the venue was prebooked for another event and had to be vacated on 7 December 1990. The inability to strike a deal on agriculture meant that the entire Uruguay Round package of agreements lapsed. Ingersent, Rayner, and Hine (1994, 73) conclude that "The crux of the lack of agreement on agriculture was that the US and the Cairns Group were unable to accept the EC's refusal to offer specific quantitative commitments to lowering border protection and reducing export assistance." *Agra Europe* (7 December 1990, P3, P5) reported that the EC was, at the last minute, willing to offer concessions in three areas. This would have involved the removal of oilseeds from the rebalancing proposal, a minimum import access arrangement for up to 3 percent of consumption, and a commitment to limit the volume of exports on which export subsidies were paid. The French minister "later disassociated himself from the concessions." At the Agriculture Council

TABLE 4.1. U.S. and EC "Offers"

United States	European Community
July 1987 • "zero option": the elimination of all farm support by 2000	*October 1987* • short-term action to balance world markets • reduction in support
October 1989 • "zero option" retained over ten years • tariffication • eliminate export subsidies over five years	*December 1989* • unspecified reduction in support • partial tariffication • tariffs and export subsidies reduced as a consequence of reduced internal support
October 1990 • tariffication, and 75 percent tariff cuts • 75 percent reduction in internal support • 90 percent reduction in export subsidies	*November 1990* • tariffication with rebalancing • 30 percent cut in aggregate measurement of support, 1986 to 1995 • no specific commitments on exports

in January 1991, Farm Commissioner Ray Mac Sharry was accused by the French minister Mermaz of having "exceeded the terms of his mandate," agreed upon after tortuous debate in November. But John Gummer countered that "the Commission must be given sufficient flexibility to negotiate" (*Agra Europe,* 25 January 1991, E2).

The EC's GATT Offer of November 1990

Although, as we have noted, the EC's position appeared to have shifted slightly at the Heysel meeting, the document of 7 November 1990 remained its formal offer. In Eurospeak, the "substance" of the offer was outlined as follows:

> The Community is prepared to reduce its support and protection by 30 percent for main products. More specifically, this implies:
>
> —a reduction of support by 30 percent, expressed by an Aggregate Measure of Support (AMS) . . . ;
> —the tariffication of certain border measures and a concomitant reduction of the fixed component resulting therefrom, together with a corrective factor; the tariffication being subject to rebalancing
> —a concomitant adjustment of export restitutions.
>
> For other products, for which the calculation of an Aggregate Measure of Support is not practicable, specific commitments will be taken. (European Community 1990, 1)

Note that tariffication was conditional upon rebalancing and that the reductions in tariff equivalents and export refunds would follow from the reduction in support. The 30 percent offer applied to main products: namely, cereals and rice, olive oil, oilseeds and protein crops, sugar beets, and livestock products (see table 4.2).

We have already noted that by backdating its offer to 1986, the EC had already achieved part of the 30 percent cut. Indeed, for wine, where the offer amounted to only 10 percent, the total AMS in 1990 was less than the target for 1995, implying that support could be increased in the interval (Annex 1B).[11] Furthermore, in figure 4.1 we demonstrated that a

---

11. The 10 percent offer also applied to fruit and vegetables, tobacco, hops, seeds, cotton, hemp, fibre flax, and silkworms.

TABLE 4.2. The EC's November 1990 Offer of a 30 Percent Reduction in Internal Support (million commercial ecu)

	Total AMS in 1986	Total AMS in 1990	Offered total AMS for 1995 (1986 level less 30 percent)	Annual percentage change in support to achieve target (1990 to 1995)
Cereals and rice	15,621	13,424	10,935	−4.0
Olive oil	3,450	3,170	2,415	−5.3
Oilseeds and protein crops	3,047	3,198	2,133	−7.8
Sugar beet	3,017	2,591	2,112	−4.0
Livestock	40,701	36,227	28,491	−4.7

Source: Annex 1A of European Community (1990). Note that in the original source the acronym *SMU* (support measurement unit) is used rather than AMS. For comment on the use of commercial ecu see the discussion in chapter 3.

Note: AMS = Aggregate Measurement of Support.

reduction in the AMS could be achieved through either a fall in volume or a cut in price and that a percentage cut in the AMS would translate into a much smaller percentage cut in price. From table 4.2 it can be seen that a good deal of compensation would have been possible within the five product groups declared. For example, by lumping all livestock products together, a further cutback in milk quotas could sufficiently reduce the livestock AMS, leaving, say, pig producers unaffected.

As earlier EC proposals had suggested, tariffication involved the determination of a fixed component that was to be subject to a "concomitant reduction" in line with the change in the AMS as well as a corrective factor. The corrective factor was very similar to the variable import levy. Annex IV of the EC's offer made clear that the corrective factor would absorb all exchange rate change against the ecu and some of any additional change in the world market price (at constant exchange rates) compared with the world market price of the reference period 1986–88 (known as the external reference price of the base period).[12] If world market prices, stripped of exchange rate variations, were to vary by up to 30 percent from this external reference price, then 30 percent of this divergence would be compensated by the corrective factor acting like a variable import levy. If the world price moved by more than 30 percent from the external reference price, then all of the additional movement would be compensated for by the corrective factor. Figure 4.2 illustrates the effect of the corrective factor if world prices, stripped of exchange rate change, were to fall below the 1986–88 external reference price. Up to 21 percent of the price decline (i.e., 70 percent of 30 percent) would be uncompensated, but otherwise the corrective factor would compensate for the price fall, acting in effect as a variable import levy.

The proposal on tariffication did, however, involve a diminution in EC preference. Under the existing CAP, variable import levies were based on the difference between a representative world market price and a threshold price. The tariff equivalent was to be based on the difference between the external reference price for the period 1986–88 "and the average Community support price (in most cases the intervention price) increased by 10 percent for the same period" (European Community 1990, 3). For common wheat, for example, the average variable import levy for the period 1986–88 was 195.7 commercial ecu, whereas the tariff equivalent had been calculated at 148.5 commercial ecu (Annex II).

---

12. See also *Agra Europe*, 12 October 1990, E6.

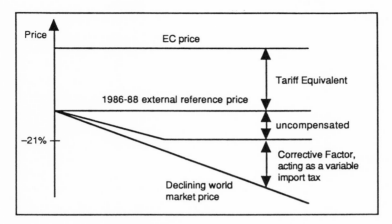

Figure 4.2. Tariffication and the corrective factor

Reprieve

Some might have thought that the December 1990 debacle in Brussels would lead to a collapse of GATT and a breakdown in world trade relations. However, the system staggered on, and the world was soon preoccupied with a real war in the Persian Gulf rather than a trade war between partners. It would have been unseemly for the allies to display publicly their trade differences at that time. In the United States, an extension of the fast-track negotiating mandate was secured, and, in the EC, discussion of the Mac Sharry reforms got underway.

CHAPTER 5

CAP Reform

In June 1984 at Fontainbleau, the European Council decided to increase the budgetary resources of the EC effective 1 January 1986, the date of accession of Portugal and Spain. In particular, the maximum VAT contribution that member states could be required to make to the EC budget was increased from 1 to 1.4 percentage points. However, in 1984 and 1985, before the Iberian Enlargement, these budgetary resources had already been committed, with an overspend on the CAP a major part of the problem. The Commission subsequently admitted that in 1985 a VAT contribution of 1.4 percentage points would have been required to balance the budget had it not been for advances made by the member states and some creative accounting that deferred expenditure to later years. In 1986 and 1987, the VAT rates would have needed to be 1.60 and 1.65 percentage points, respectively (Commission 1987b, 4).

A curious feature of the accounting procedures for intervention stocks at the time was that stocks were valued at the intervention price, not at a realistic sale price. Furthermore, they remained in the ownership of the member states—EC budget funds had not been used in their purchase. By allowing intervention stocks to rise, the Commission—with the full knowledge of the member states—had in effect deferred expenditure to later years because, when intervention stocks were eventually sold off at a loss, this would trigger an enormous budgetary charge against the EC. This financial overhang was referred to as the legacy of the past (in the EC jargon of the day). In March 1987, in an attempt to liquidate the excessive intervention stocks of butter, the Council decided to sell off stocks in 1987 and 1988 with the accounting "loss" to the member states being reimbursed not from the 1987 and 1988 budgets but rather from the 1989–92 budgets.[1] Much of this butter was exported and subsequently became part of the base quantity of exports recorded in the period 1986–90 for the purpose of administering the export constraints

---

1. Council Regulation (EEC) No. 801/87, *Official Journal of the European Communities* L79, 21 March 1987.

negotiated in the Uruguay Round. The Commission (1989, 43) subsequently reported that a "significant part" of the export sales were of "old butter" at prices "well under 10 percent of the intervention price."

As shown in table 5.1, during the mid-1980s the recorded budgetary costs of the CAP continued to rise despite the introduction of guarantee thresholds in 1982 and milk quotas in 1984, as outlined at the end of chapter 3. In part, this escalation was linked with the falling value of the U.S. dollar against European currencies (as shown in figure 2.3). World prices tend to be sticky in dollar terms, and thus a falling dollar means lower world market prices when converted to ecu and higher per unit export refunds under the CAP. It was against this background that in July 1987 the European Council "confirmed once again . . . the urgent need to trim supply down to demand levels by action as a result of which the market would gain more influence as a stabilizing force" and asked the Commission "to produce appropriate proposals" (Commission 1989, 15). Would the rhetoric result in substantial policy change on this occasion?

Agricultural Stabilizers and Budgetary Discipline

The package deal that emerged at the meeting of the European Council in February 1988 involved a further extension of the budgetary resources made available to the EC to fund both the increasing range of policies it pursued, and what was declared to be fundamental reform of the CAP. Indeed, Britain's prime minister, Margaret Thatcher, was known to have pressed for fundamental reform of the CAP in exchange for Britain's assent to an expansion of the budgetary revenues of the EC. *The Observer* (14 February 1988, p. 11) reported that

> It was an exhausted and only half convinced Mrs Thatcher who presented the results of the European Summit in Brussels to a sceptical Press corps at 2 A.M. yesterday morning. Against all expectations, and in defiance of the Prime Minister's most deeply entrenched instincts, the Summit had ended in success.

What had happened was that once again a form of words had been found to suggest that fundamental reform of the CAP was underway, whereas in truth the constraints on the excesses of farm support were limited.

The budgetary discipline for the CAP was expressed in a commitment that automatic agricultural stabilizers would come into force if the annual growth in budgetary expenditure on CAP price support exceeded 74 percent of the increase in the EC's gross domestic product (GDP).

TABLE 5.1. Budget Expenditure on the CAP, 1982–88 (millions of ecu)

	1982	1983	1984	1985	1986	1987	1988
FEOGA[a] Guarantee	12,406	15,812	18,346	19,744	22,137	22,968	27,687
FEOGA Guidance	650	728	676	720	774	907	1,143
Total Budget	20,706	24,808	27,209	28,085	35,174	35,469	41,121
Guarantee as a percentage of total	59.9	63.7	67.4	70.3	62.9	64.8	67.3

Source: Commission of the European Communities, *The Agricultural Situation in the Community, 19— Report*, various years.

Note: The accession of Portugal and Spain in 1986 affects these figures.

[a] FEOGA is the French acronym for the European Agricultural Guidance and Guarantee Fund. This is not a separate fund but the name given to a particular section of the EC's general budget. The guarantee section funds price support, whereas the guidance section funds structural policy.

The Commission estimated that this would restrict the growth in expenditure on the CAP to 2 percent per year in real terms, whereas between 1980 and 1987 the corresponding figure had been 6 percent. The Commission (1989, 16) fatuously declared: "it is clear that the new rule is a draconian one."

Details varied from sector to sector, but the basic idea was that the Council had ceded authority to the Commission to apply automatic cuts in support prices if certain thresholds were exceeded. For some sectors these were expressed as maximum guarantee quantities, virtually reinventing the guarantee threshold that had been abandoned for cereals in 1986. For oilseeds, depending upon the actual harvest, the cuts could be substantial, but for cereals they were limited to 3 percent per year and involved a tremendously complex system of "additional" coresponsibility levies.[2] A voluntary set-aside scheme was also introduced, which attracted little take-up.

Despite drought in North America in 1988 that resulted in a substantial increase in world market prices and a corresponding reduction in export refunds, the EC produced a bumper cereal crop that triggered the stabilizer mechanism in 1989. But despite the European Council's determination, as expressed at its meeting in Brussels in February 1988, to apply draconian controls to the CAP, the EC Council of Agriculture Ministers failed to show the same resolve. At the 1989–90 annual fixing of CAP prices, ministers tried to relieve the pressure on the farm population as best they could. Manegold (1989, 45) writes of Germany "tenaciously" fighting "for every tenth of a percentage point in the monetary gap which . . . allowed some national support prices to be kept marginally higher than originally proposed by the Commission" and concludes, "Thus, the much-touted CAP reform has perhaps ended before it really got to the core."

The Mac Sharry "Reforms"

It seems clear that the Mac Sharry reform proposals had their origin in the internecine struggles of October 1990, as the EC sought to produce a GATT offer to meet its international obligations. To cajole the ministers, particularly the German representative, to accept the proposed 30 percent reduction in support, the Commission had to concede that compensation would be paid to farmers "in the form of direct income supports that do not stimulate production." Nonetheless, the Mac Sharry

---

2. For details see, for example, Manegold (1989).

reforms were launched onto an unsuspecting world, and despite the obvious need to engineer some changes to the CAP if any form of GATT accord were to be agreed upon—let alone put in place the EC's own GATT offer—the official fiction was maintained for many months that the Mac Sharry reforms and GATT talks were two unrelated policy initiatives.

On 11 January 1991, *Agra Europe* (11 January 1991, P2) reported that the Commission had begun work on a new reform plan. The report emphasized the escalating budgetary cost of the CAP and indicated that this was the prompt for the Commission's thinking. Indeed, the report stated authoritatively: "What is certain is that the Commission's budget-driven internal reforms have very little to do with the current GATT negotiations." Later, Ray Mac Sharry was reportedly concerned to identify the source of the premature leaks of his reform plan. We can only speculate on who leaked what and why, but a Machiavellian view might be that the source of the leak was close to Mac Sharry and that the news agencies had been fed the story that these were budget-driven, internal reforms unconnected with GATT. As the months progressed and the plans crystallized, it became clear that, rather than reduce EC budget spending, the reform process was likely to increase expenditure.

Agra Europe continued to report leaked documents, and, in particular, on 18 January 1991, the full text that had apparently been submitted to the commissioners by the directorate-general for agriculture. However, it was not until 1 February 1991 that the formal document, stripped of its numerical detail, was published and submitted to the Council (Commission 1991a).

The document stressed the problems faced by the CAP, without directly mentioning GATT. Thus, it pointed out that "between 1973 and 1988 the volume of agricultural production in the EEC increased at 2 percent per annum whereas internal consumption grew by 0.5 percent only per annum," resulting in an accumulation of stocks and an increase in the EC's exports onto a stagnant world market, which "goes some way towards explaining the tension between the Community and its trading partners." It encouraged intensive farming systems, giving rise to environmental concerns. It was an asymmetric policy that failed to recognize the significant dispersion in farm sizes throughout the EC: "80 percent of the support . . . is devoted to 20 percent of farms which account also for the greater part of the land used in agriculture." Despite ever increasing budget costs, "The per capita purchasing power of those engaged in agriculture has improved very little over the period 1975–89" (Commission 1991a, 1–3). Formal proposals, detailing the changes to be made, were not published until July 1991 (Commission 1991b).

Cereals, Area Compensatory Payments, and Set-Aside

At the heart of the Mac Sharry plan were the proposals for the arable sector. The basic elements were as follows:

- levels of price support in the cereals sector to be reduced substantially, bringing them much closer to world market levels;
- farmers to be compensated for their loss of revenue through a system of area payments;
- the compensation, however, to be modulated; that is, "small" farmers would be compensated in full, but beyond a certain size only partial compensation would be paid; and
- compensation to be linked to a set-aside scheme.

There were three obvious inconsistencies with the EC's GATT offer. First, the numbers were different. Thus, COM(91)258 (Commission 1991b, 9) would set a "target price" (clearly a different concept from the previously existing target price in the cereals sector) some 35 percent "below the existing average buying-in price for cereals," and the new intervention price would be some 10 percent below that.

Second, the Mac Sharry plan made no concession to the concept of tariffication: indeed, COM(91)258 states quite clearly that "The basic principles and instruments of the common market organisation for cereals will be maintained" (Commission 1991b, 9). Finally, the proposed compensation payments could hardly be described objectively as "decoupled" and thus eligible for inclusion in the green box. Indeed, if the Mac Sharry reform were to go ahead and subsequently in the GATT talks it were to be decided that the compensatory payments should be classified in the amber box, a further reform of the CAP would be necessary simply to meet the EC's GATT offer. However, if a political deal could be hatched with the Americans and the compensatory payments (and the Americans' deficiency payments) classified in the green box, then the Mac Sharry reforms were more than was required to meet the EC's GATT offer. These thoughts did not escape the notice of skeptical ministers.

Economists might question the need to compensate farmers at all. In other industries, businesses do not expect the government to compensate them for faulty investment decisions; why should farming be different? Politically, however, some degree of compensation is probably essential. The question then became, who should be compensated and how?

As the discussions progressed, it became obvious that modulation of the compensation was not politically acceptable, and thus by the time the package was agreed upon in May 1992 large farmers were not to be penalized at all. This presumably still left 20 percent of producers collecting 80 percent of the support. The evolution of the concept of modulation is summarized in table 5.2.

According to the leaked proposals, the intervention price for cereals would have been reduced to 90 ecu per metric ton, with a target price set at 100 ecu per metric ton. A system of direct area payments would have been introduced "calculated in such a way as to compensate (on average) the total loss of income for farms with up to 30 hectares of cereals." In this context, "cereals" seems to mean cereals, oilseeds, and protein crops. The compensatory payments would have been reduced by 25 percent on the next 50 hectares of cultivated crops and by 35 percent on all "cultivated land over the first 80 hectares." Eligibility for these compensatory payments would have been dependent upon farmers setting aside a portion of their land: that is, nothing for the first 30 hectares, 25 percent for the next 50 hectares, and 35 percent "for areas over the first 80 hectares." This, the Commission pointed out, meant that "a farmer with 50 hectares would have to set aside 10 percent of his land as temporary fallow and a farmer with 100 hectares 19.5 percent" (*Agra Europe*, 18 January 1991, E2). It was not clear whether compensation would be paid on set-aside land, but the Commission claimed that "The targeted modulation of the aid scheme will provide an effective response to the enormous social gap between the 4m small farms with less than 30 hectares of cereals and the 54,000 farms with more than 50 hectares (1.3 percent of farms) who occupy 22 percent of the surface and account for around one third of production" (*Agra Europe*, 18 January 1991, E4). Sir Simon Gourlay, president of the National Farmers' Union of England and Wales, commented: "The ideas in the leaked Commission paper would devastate British farming" (*Agra Europe*, 18 January 1991, E8). Britain's minister of agriculture, John Gummer, embarked upon an energetic campaign to oppose modulation.[3]

When the Commission announced its formal proposals in July 1991, in COM(91)258, the Mac Sharry plan for reform of the cereals regime had been softened in three important respects. First, the area compensatory payments would be paid in full rather than being reduced on a per hectare basis as farm size increased. Second, the set-aside requirement had been reduced to 15 percent (for rotational set-aside) of the eligible arable land. Under the leaked proposal, in order to claim compensatory

---

3. See, for example, his speech to the National Farmers' Union Annual General Meeting in February 1991 (Gummer 1991).

TABLE 5.2. *Modulation* and the Mac Sharry Plan

	Leaked text January 1991[a]	Proposed in COM(91)285/3	Agreed upon May 1992
Area Compensatory Payments: Paid in full on 30 hectares, at 75 percent on next 50 hectares, at 65 percent on remainder		Paid in full	Paid in full
Set-Aside Obligation: 0 percent on 30 hectares, 25 percent on next 50 hectares, 35 percent on additional land		0 percent under the "small farmer" scheme (about 20 hectares), otherwise 15 percent	0 percent under the "small farmer" scheme (about 20 hectares), otherwise 15 percent
Compensation for Set-Aside: Uncertain		At cereals rate for areas up to about 7.5 hectares	At the cereals rate for all set-aside land. Later increased by 27 percent

[a] As reported in *Agra Europe* (18 January 1991, E1).

payments, a farmer with 100 hectares had to set aside 19.5 hectares. Third, area compensatory payments would be paid on a maximum area of set-aside land per farmer, reckoned to be on average about 7.5 hectares in the EC.[4]

When the Mac Sharry reforms were finally adopted in May 1992 the concept of modulation had totally disappeared. Farmers could claim area compensatory payments on unlimited areas of cropped and set-aside land. Furthermore, in the 1993–94 farm price settlement, set-aside payments were increased by 27 percent. All claims that the Mac Sharry reforms would be targeted in favor of small farmers and that the new CAP would be less costly to the EC's budget than its predecessor had by now been long forgotten.

The Mac Sharry package did, however, fundamentally change the hierarchy of cereal pricing under the CAP as well as the method of support. The price changes are shown in table 5.3. In Regulation (EEC) No. 1766/92, the Council fixed July target, threshold, and intervention prices for all cereals for the 1993–94, 1994–95, 1995–96, and all subsequent marketing years (Council 1992).[5] This was a major change to the institutional framework within which CAP policy is determined. Previously, cereal support prices had been fixed by the Council annually on a proposal from the Commission. The balance of power lay with the Council. Now support prices are fixed forever and can only be changed by the Council if and when the Commission forwards a proposal to the Council. The balance of power has shifted in favor of the Commission. Consequently, the annual farm price reviews will have fewer items on the agenda, and cereal pricing will be less susceptible to horse-trading and last-minute package deals.

The 1995–96 target price of 110 ecu per metric ton was 29 percent lower than the average buying-in price of 155 ecu per metric ton in 1991–92. A more substantial price cut of 35 percent had been proposed, which would have brought the target price down to 100 ecu per metric ton. In

---

4. The area would have been determined on a regional basis. It was set equal to the set-aside obligation (i.e., 15 percent of the area needed to produce 230 metric tons of cereal at the regional yield). For the EC on average, this was thought to be 15 percent of 50 hectares (Commission 1991b, 13). The "small producer" cutoff was determined in a similar fashion: the area needed to produce 92 metric tons at the regional yield and said to be about 20 hectares on average (15.51 hectares in England).

5. Note that as a consequence of the green conversion rate regulations outlined in chapter 3, while the provisions regarding the switchover mechanism remained in force, these ecu prices were to be reduced at the beginning of each marketing year by the Commission, acting under its own authority, to remove 25 percent of any subsequent increase in the switchover coefficient determining the value of the green ecu.

addition, an intervention price for all eligible cereals of 10 ecu per metric ton below the target price was also set.[6]

In the original proposal, the threshold price would have been 10 ecu per metric ton above the target price, considerably reducing the degree of community preference incorporated into the regime. In the May 1992 package, the threshold price was set at 155 ecu per metric ton, reflecting the member states' opposition to the original proposal. Because the threshold price is particularly important for determining market prices in such circumstances, this change was of considerable interest to EC producers of quality grains (for example bread-making wheat or malting barley) as well as to producers in grain-deficit regions.

The change from a buying-in price of 155 ecu per metric ton in 1991–92 to a target price of 110 ecu per metric ton in 1995–96 represents a price cut of 45 ecu per metric ton, and this amount is the basis for the determination of the area compensatory payments for cereal crops. Each member state had to specify a number of regions and determine the average annual yield of cereals in that region over the five-year period 1986–87 to 1990–91, excluding the years with the highest and lowest yields. The United Kingdom declared five regions, with an average cereal yield in England of 5.93 metric tons per hectare. Thus, on the basis of the figures in table 5.3, eligible cereal growers in England could

TABLE 5.3. Cereal Prices under the Mac Sharry Reform (July prices in ecu per metric ton)

	Threshold	Target	Buying-in/Intervention	Aid (Basic Amount)
1991–92	228.67	NA[a]	155[b]	NA[a]
1993–94	175	130	117	25
1994–95	165	120	108	35
1995–96 on	155	110	100	45
Proposal in COM(91)258	110	100	90	55

Source: Official Journal of the European Communities, L181, 1 July 1992, pages 15 and 23; and Commission (1991b), pp. 9–10.

Note: The threshold price quoted for 1991–92 is for common wheat. The buying-in price quoted for 1991–92 is roughly midway between that for bread wheat and feed grains.

[a] NA = not applicable.

[b] From which a coresponsibility levy was deducted.

---

6. Subsequently it was decided that wheat only suitable for animal feed was not eligible for intervention, thus causing market prices to fall below support levels for this product.

expect to receive area compensatory payments of 266.85 ecu per hectare (calculated at 45 ecu per metric ton and 5.93 metric tons per hectare) in 1995, whatever their own crop yields. Similar arrangements apply for oilseeds and protein crops. If, in any one year, the total area in any region on which compensatory payments are claimed exceeds that region's historic base, then all compensatory payments are scaled back by the same percentage and an additional element of uncompensated set-aside is introduced for the region's growers in the following year.

As table 5.2 shows, "small farmers" are exempt from a set-aside requirement, but for larger farmers receipt of the area compensatory payments is conditional upon set-aside—initially set at 15 percent for rotational set-asides. What this means is that any grower, however large, can claim area compensatory payments on the "small farmer" area without set-aside. But to claim on larger areas a grower must set aside. The rotational set-aside payment was initially set at the 1995 cereals rate but was subsequently increased and set at 57 ecu multiplied by the region's historic cereal yield.

An Unfinished Agenda

The Mac Sharry reforms were not limited to cereals, oilseeds, and protein crops. They also introduced a quota mechanism on the headage payments for sheep and extended the importance of headage payments for beef while weakening intervention arrangements for beef. However, the changes to the dairy regime were trivial compared to the proposals contained in COM(91)258 and foreshadowed in the leaked document of January 1991. In addition, there was a series of "Accompanying Measures" embracing an "Agri-Environment Action Programme," "afforestation of agricultural land," and early retirement schemes for farmers.[7]

It is important to note that both the beef and sheep premiums, like the arable area and set-aside compensatory payments, were limited to animal numbers and areas cropped at a date before the Mac Sharry reforms were enacted. As we discuss in chapter 6, this enabled the Commission to claim at Blair House that such payments are decoupled and hence exempt from GATT disciplines. This understanding was incorporated into the GATT deal concluded on 15 December 1993 in Geneva.

However, as noted above, the Mac Sharry package failed to reform the dairy sector. Furthermore, sugar, wine, olive oil, and fruits and

---

7. For an overview of CAP policy changes introduced in 1992, see Swinbank (1993a).

vegetables were untouched. In addition, the implementation of the beef headage payments soon gave rise to considerable market disruptions, generating demands for further reform. Clearly, there was, and still is, a good deal of unfinished business for the Commission and the Agriculture Council to consider.

Retrenchment

It is doubtful whether the word *reform* can properly be applied to the changes to the CAP introduced in May 1992. They were undoubtedly a significant change, but they were designed to maintain the revenues of the farm sector and to keep farmers on the land. The changes began a shift from the "old" CAP to a newer, more market-oriented model, but we suspect that a decade from now the Mac Sharry "reforms" will simply be seen as a small part of a long drawn-out reform process. We leave to chapter 7 an assessment of the effectiveness and limitations of the Mac Sharry reforms and to chapter 8 our recommendations for the reform process over the next decade. We will close this chapter, however, by noting that the farm lobby will not give up its privileged position without a struggle. Notwithstanding the generosity of the compensation terms for cereal producers included in the May 1992 package, ministers have since clawed back concessions for their farmers.

First, as we have already noted, the Council has increased the rate of aid for set-aside land by 27 percent. This was widely seen as a "sweetener" for the French farm lobby, to buy off their opposition to the Blair House Accord, which will be discussed in chapter 6 (*Agra Europe,* 28 May 1993, E1). Second, greater flexibility has been introduced into the set-aside requirements. Originally set at 15 percent on a six-year rotation, a 20 percent nonrotational scheme was also introduced, and some member states soon had in place a 20 percent scheme on a truncated rotation.[8] Set against this greater flexibility is the fact that member states impose quite onerous management requirements on set-aside land, often prompted by environmental concerns, and environmental lobbyists continue to press for the introduction of what they refer to as an element of cross compliance in the aid schemes. By this they mean that compensation payments should only be made to farmers who demonstrate a minimum attainment of environmentally friendly husbandry.

Third, the areas on which the additional aid for durum wheat pro-

---

8. The nonrotational set-aside scheme was set at 18 percent for the United Kingdom and Denmark. See Ansell and Vincent (1994) for implementation details.

ducers is paid has been increased. In the 1994–95 farm price settlement, ministers agreed to pay aid on an additional 50,000 hectares of durum wheat production in northern France, where—it was claimed—pasta makers had experienced supply difficulties (*Agra Europe,* 22 July 1994, E1)

Fourth, skeptics doubt that the rules will be applied rigorously. Experience with the application of milk quotas in the Mediterranean regions does not inspire confidence. Surprisingly, despite the commercial attractiveness of the scheme, for the 1993 harvest the level of area compensation claims seems to have been 5.6 percent lower than the feasible maximum, as shown in table 5.4.[9] Nonetheless, in some areas, there was an overshoot. In various *Länder* of the former East Germany, in parts of Spain, and in Scotland the total claims exceeded the base area. The largest overshoot was of 15 percent in Mecklenburg–Western Pomerania. In all cases, local farmers claimed that it was the base area computations that were at fault, not the levels of claims. Germany strongly resisted the application of sanctions (a proportional scaling back of payments plus an additional uncompensated set-aside requirement for the following year) against its producers. In the face of German intransigence, the Council agreed that in the East German *Länder* the penalty scheme would not be fully operational until 1996. In 1993 only 10 percent of the full rate of the penalty would apply, but there was no increase in the base area (*Agra Europe,* 15 October 1993, P3). Having conceded the point for Germany, it was difficult to resist making similar concessions to other regions that had overclaimed, and in September 1994 member states facing an overshoot on their 1994 claims cited the precedent of the preceding year (*Agra Europe,* 9 September 1994, E6).

Finally, it cannot be said that the December 1992 decisions on revisions to the green money system, outlined in chapter 3, strengthened the Council's resolve to hold the line. Germany was again at the forefront in arguing that the revised arrangements for green money should not trig-

---

9. Despite its complexity, table 5.4 has been included to demonstrate the diversity of experience across the Union. It sets out for each member state as well as for the EU, first, the total base area on which claims can be made and, second, the claims actually made in 1993–94. These are listed as claims made under the small producer scheme, those under the professional producer scheme to which the set-aside requirement is attached, residual claims under the now defunct voluntary five-year set-aside scheme introduced in 1988, and a deduction for forage areas to avoid double payments on the same area of land under both the area compensation and livestock headage payment schemes. The final row shows the total area of set-aside, in both schemes, expressed as a percentage of the total area of claims in 1993–94.

TABLE 5.4. EC Area Compensation Claims, 1993–94 (1,000 hectares)

	UK	Portugal	Nether-lands	Luxem-bourg	Italy	Ireland	France	Spain	Greece	Germany	Denmark	Belgium	EC-12
Base area	4,407	1,054	436	43	5,800	345	13,522	9,229	1,491	10,002	2,017	479	48,825
Small producer claims	314	250	282	24	2,409	117	2,511	2,670	1,163	2,236	477	278	12,731
Large producer claims	3,664	433	51	11	1,293	153	10,409	5,765	104	7,091	1,328	132	30,434
Total claims	3,978	683	333	35	3,702	270	12,920	8,435	1,267	9,327	1,805	410	43,165
5-year set-aside	133	0	15	0	639	2	214	95	1	415	7	1	1,522
Forage area	135	130	23	3	130	25	327	72	20	385	100	50	1,400
Total area	4,246	813	371	38	4,471	297	13,461	8,602	1,288	10,127	1,912	461	46,087
Overshoot	-161	-241	-65	-5	-1,329	-48	-61	-627	-203	125	-105	-18	-2,738
Percentage overshoot	-3.7	-22.9	-14.9	-11.6	-22.9	-13.9	-0.5	-6.8	-13.6	1.2	-5.2	-3.8	-5.6
Large Producer Claims:													
Cereals	2,511	266	40	7	798	122	6,699	3,175	76	4,871	862	103	19,530
Oilseeds	371	87	2	1	252	2	1,380	1,660	12	1,074	143	3	4,987

Protein crops	226	2	2	1	36	5	741	22	0	83	117	6	1,241
Set-aside	556	78	8	2	207	24	1,589	909	17	1,063	205	19	4,677
Total	3,664	433	52	11	1,293	153	10,409	5,766	105	7,091	1,327	131	30,435
Set-aside as percentage of total	15.2	18.0	15.4	18.2	16.0	15.7	15.3	15.8	16.2	15.0	15.4	14.5	15.4
"Small," as percentage of "large"	8.5	58.0	542.3	218.2	186.3	75.8	24.1	46.3	1,106.7	31.5	35.9	212.2	41.8
Total Claims on Cropped Area:													
Cereals	2,821	500	321	31	3,207	238	9,184	5,368	1,232	7,084	1,339	380	31,705
Oilseeds	372	103	2	1	252	2	1,394	2,128	17	1,091	143	3	5,508
Protein crops	228	3	3	1	36	5	753	31	1	89	117	7	1,274
Total cropped area	3,421	606	326	33	3,495	245	11,331	7,527	1,250	8,264	1,599	390	38,487
Set-aside as percentage of total	16.2	9.6	6.2	5.3	18.9	8.8	13.4	11.7	1.4	14.6	11.1	4.3	13.5

Source: European Commission, as reported in *Agra Europe* (18 February 1994, E1), with an amendment for Greece.

ger a revaluation of the green mark, resulting in a reduction in deutsche mark prices in Germany. Thus, the switchover mechanism was retained for a further two-year period, and in fact the switchover coefficient did increase in 1993. Ironically, it was a German presidency that presided over the Council's deliberations on a renewal of the system at the end of 1994.

When the European Council met in Edinburgh in December 1992, the troubled times experienced by the EC resulted in a crowded agenda. The Danish electorate had recently rejected the Maastricht Treaty in that country's first referendum on the question, the French electorate had only narrowly accepted it, and considerable disquiet had been expressed elsewhere. The pound sterling had in September ignominiously exited from the exchange rate mechanism (ERM), raising doubts about the viability of monetary union. Among other decisions, the heads of state or government reached agreement on financing the EC budget over the period 1993–99, which became known as the Delors II Package. They reiterated their decision of February 1988, made in Luxembourg, that the growth of expenditure on agricultural price support should be limited to 74 percent of the growth in EC GDP but went on to note that "recent monetary movements will result in a significant increase in FEOGA Guarantee expenditure." This conclusion led them to agree to a weakening of the financial guidelines agreed upon for farm spending, because they concluded that "if, as a consequence of this increase, agricultural expenditure were to exceed the guideline and thus compromise the funding of the new Common Agricultural Policy as already approved, appropriate steps to fund the FEOGA Guarantee will be taken by the Council" (European Council 1992, 66). Is it any wonder that Europe's trading partners remained suspicious of the EC's real intent?

CHAPTER 6

The Negotiations II: From the Draft Final Act to Marrakesh

The breakdown in December 1990 of the Brussels talks, which had been convened to conclude the Uruguay Round, did not end the process. Although hostilities in the Persian Gulf kept most American diplomats in Washington, the GATT director-general doggedly continued his shuttle diplomacy and kept the hope of a deal alive. In February 1991, at an emergency meeting in Geneva, all the participants "agreed to the resumption of negotiations on the basis of reaching 'specific binding commitments to reduce farm supports in each of three areas: internal assistance, border protection and export assistance'" (Ingersent, Rayner, and Hine 1994, 78). Ingersent et al. conclude that "By not objecting to this form of words, the EC softened its previous hardline refusal even to consider the possibility of acceding to the US demand for specific support reduction commitments in all three areas" (78).

Otherwise, there was little progress in 1991. In the United States, Edward Madigan succeeded Clayton Yeutter as secretary of agriculture, and Congress extended the fast-track negotiating mandate for a further two years. Hillman (1994, 50) reports that the new secretary of agriculture immediately faced opposition from the American dairy lobby over the Bush administration's hard-line stance on trade liberalization in the Uruguay Round. In Hillman's view, the outcome of this confrontation was that "Other farm lobbyists and farm groups hostile to the Administration in the GATT had won a victory of sorts and would inevitably expect a softening of the hard line. The zero option was history."

The Draft Final Act

Meanwhile GATT Director-General Arthur Dunkel worked on a compromise plan. On 20 December 1991, he tabled a Draft Final Act setting out a comprehensive agreement covering all the issues under discussion in the Uruguay Round, drawing on such consensus as existed in the various negotiating groups, in the hope that this package deal would form the basis of the elusive final agreement (GATT 1991). The text was substantial, with the chapter on agriculture alone amounting to seventy-four pages.

Dunkel's hope was that an agreement closely modeled on this document could be secured in 1992. However, the initial response from the EC was hostile. On 23 December 1991, a meeting of the EC's Foreign Trade Council, which some agriculture ministers attended, rejected the Dunkel draft, "saying portions of it will have to be negotiated" (*Agra Europe,* 3 January 1992, E3). The Dutch president of the Council was reported as saying, "A text which does not take into account the principles of the reform of the CAP is not acceptable" (E3).

The agriculture chapter of the Draft Final Act had clearly evolved from the earlier discussions and quasi agreements that had emerged in the Agriculture Group in that it set out specific commitments concerning import access, domestic support, and export competition, and in addition there was to be an agreement on sanitary and phytosanitary matters. GATT signatories were being asked to "tariffy" all border measures, reduce tariffs by 36 percent on average over a six-year period, cut domestic support by 20 percent, trim expenditure on export subsidies by 36 percent, and reduce the volume of subsidized exports by 24 percent over the same six-year period. Despite the EC's initial response, these basic proposals are clearly discernible in the final agreement of 15 December 1993, and so consideration of the detail of the proposal will be deferred until later in this chapter.

Although in the early weeks of 1992 an agreement looked as elusive as ever, the Draft Final Act did specify that by 1 March 1992 countries should submit details showing how the import access, domestic support, and export competition commitments outlined in the Draft Final Act would be implemented, so that final tariff schedules and other binding commitments could be established by 31 March 1992. Predictably, this ruffled feathers in the EC, and once again the EC was unable to respect the set deadline. On 2 March 1992, the Commission, after Council discussions but without a Council mandate, submitted the technical material but not the commitments to reductions in support asked for by Dunkel. The French government was particularly insistent that in submitting this document the EC did "not accept the Dunkel paper of December 1991, nor any other of his conclusions that would alter the EC negotiating position in Geneva" (*Agra Europe,* 6 March 1992, E13).

The Oilseeds Dispute

Cynics might suggest that it was only the threat of a trade war that brought the EC to the negotiating table in Washington in November

1992. Over the intervening months, the lack of an agreement on agriculture had seemed to be the most substantial element blocking conclusion of the GATT round, and the EC was seen as one of the most influential participants resisting agreement on agriculture. But even within the EC it was realized that the Draft Final Act would have to serve as the basis for a final accord. The chief protagonists in this impasse were now the EC and the United States: Japan and others with protected agricultural sectors kept their heads down, and while Australia continued its lobbying activities the Cairns Group itself largely vanished from the scene. A deal on agriculture would have to be brokered between the EC and the United States, and, until it was, the GATT negotiations in Geneva were on ice.

Oilseeds and a pending presidential election in the United States forced the next step in the saga. In the Dillon Round in the early 1960s, as discussed in chapter 1, the EC had entered into GATT bindings on oilseeds, which meant that products like soybeans could enter the EC duty-free. With similar GATT bindings on manioc, EC animal feed compounders had found that a mixture of carbohydrates (such as manioc) and vegetable proteins (such as soybean meal) were a perfectly acceptable substitute for cereals in their manufactured animal feeds. And in the Alice-in-Wonderland world of the CAP these "cereal substitutes" were cheaper than cereals. Understandably, their use expanded, and increasing quantities of soybeans were shipped across the Atlantic from growers in the United States to crushers in the EC so that soybean meal could be incorporated into animal feeds.

However, in American eyes, this lucrative trade was threatened in the 1970s when the EC introduced a support system for oilseeds under the CAP. Because import levies or tariffs were excluded by the GATT binding, the EC chose to support oilseed producers through direct subsidy payments. In the main, these subsidies were paid to crushers on the understanding that the crushers paid an enhanced price, reflecting the subsidy, to growers. As world oilseed prices fell, the subsidies were increased, and vice versa, in order to maintain the level of prices received by EC growers.[1]

The United States argued that these subsidies encouraged EC production of oilseeds and hence had an adverse impact on the volume of U.S. soybeans imported into the EC—thereby negating the Dillon Round GATT binding entered into by the EC. Two GATT arbitration panel rulings upheld this view. The EC's plan had been to resolve this issue in the context of an overall GATT agreement. This in part was why

---

1. See Harris, Swinbank, and Wilkinson (1983, 138–41), for details.

the Europeans had been seeking rebalancing. The EC had wanted to negotiate a release from the GATT bindings, allowing it to impose tariffs or levies on oilseeds in exchange for a reduced level of support for cereals. But the Americans had rejected rebalancing, the Draft Final Act did not embrace this concept, and the negotiations had not been concluded. The United States was determined to take its revenge.

Furthermore, the Bush administration was eager to secure a GATT deal before the presidential elections of November 1992. Faced with the impasse over oilseeds and the collapse of yet another round of talks between the two parties on 3 November 1992, the United States threatened to introduce punitive import tariffs in December 1992 on a range of agricultural and food products, carefully selected to target the EC in general and France in particular (*Financial Times*, 7 November 1992, p. 1). The drama was heightened two days later when the EC's chief agricultural negotiator in the GATT talks, Commissioner for Agriculture Ray Mac Sharry, resigned his brief as GATT negotiator (but not as farm commissioner), accusing EC Commission President Jacques Delors of undermining his position and "of betrayal of trust" (*Financial Times*, 7 November 1992, p. 1). Mac Sharry, however, resumed his role and returned to the United States later that month.

The Blair House Accord

Blair House, the president's official guest house on Pennsylvania Avenue, was the venue for the talks. The deal, first referred to as an "agreement," then later as an "accord," and by the French as a "pré-accord," was concluded on the evening of Friday, 20 November 1992, some seventeen days after the American electors had failed to reelect George Bush to the White House. Three things were "settled" at Blair House. First, a bilateral agreement between the EC and the United States was established that resolved the oilseeds dispute, which—although a GATT issue—was not strictly speaking part of the Uruguay Round. (It has, however, subsequently become part of the EC's commitments entered into on 15 December 1993.) Second, another bilateral agreement covering maize gluten feed was reached. And, third, a series of understandings on how some articles of the agriculture chapter of the Draft Final Act should be amended were agreed upon.

It was an agreement between the outgoing American administration and the EC Commission. In Paris, the French prime minister, Pierre Beregovoy, told the National Assembly that the Commission had exceeded its negotiating mandate granted by the Council in 1990, that the Blair House Accord was therefore unacceptable to France, and that

France "will use its power of veto under the Luxembourg compromise if the French position is not heard and respected" (*Agra Europe,* 27 November 1992, E4). Beregovoy argued that France "has the right to expect from its friends the solidarity it has never haggled over with them"—in other words, that France should be allowed to veto the deal. The National Assembly was told that if Belgium, Italy, and Spain were to vote with France, a qualified majority vote in favor of the deal could not be secured (*Agra Europe,* 27 November 1992, E5).

In Washington, Ray Mac Sharry insisted at a press conference called immediately after the Blair House Accord had been struck that the accord was compatible with the CAP reforms agreed upon by the EC earlier in the year and that it was "not a disadvantage for EC farming, but an advantage because it consolidates the reform of the CAP internationally" (*Agra Europe,* 27 November 1992, E1). In the weeks that followed, the debate over the compatibility, or lack of it, of the Blair House Accord was vigorously pursued in farm policy circles in Europe (these issues will be outlined later in chapter 7).

Meanwhile, in Geneva, a meeting of the Trade Negotiating Committee agreed that "substantive negotiations on multilateral issues should begin immediately" in the hope that an overall GATT agreement could be put in place by the year's end (*Agra Europe,* 27 November 1992, E4). This overly optimistic view was, however, soon punctured when it became evident that France was firmly opposed to the deal and that President Bill Clinton was in no hurry to adopt what was initially seen as a flawed agreement entered into by his predecessor.

But what did the Blair House Accord amount to? It was not easy for "outsiders" to gain access to the details, other than to read accounts of what had been agreed upon in the press. Neither the American administration nor the EC Commission, as far as we can tell, published an official text of the accord. However, the Commission did quickly publish its explanation of what had been agreed to (Commission 1992), the USDA published an account of the accord (Herlihy, Glauber, and Vertrees 1993), and late in 1993 the Commission made available an undated and unattributed document entitled *Changes to the Draft Final Act Required by the U.S./EC Blair House Agreement* for persistent inquirers.[2] The lack of any documentation clouded the issues and meant that some of the debate in Europe over the acceptability of the accord was based upon hearsay rather than fact. An example of the ensuing

---

2. The text of the *Memorandum of Understanding on Oilseeds* was, however, published as an appendix to Commission (1993b).

confusion is the erroneous suggestion in the French government's memorandum of August 1993 attacking the Blair House Accord that food aid shipments would be subject to limitations (*Agra Europe,* 3 September 1993, E1).

A French Veto?

Within the EC there was much talk of a French veto, but it soon became evident that the Commission would not put the U.S.–EC farm agreement to the Council for formal approval. The plan was that, unless challenged, the Commission would continue with the negotiations in Geneva and would only present to the Council the overall package, which subsumed agriculture, sometime in 1993.[3] The presumption was that in the Council a qualified majority would approve the deal and that all member states would then, as GATT signatories, endorse the agreement in Geneva.

In January 1993 the French position had significantly weakened. In December 1992, Ignaz Kiechle—who had staunchly served as Germany's minister for agriculture since 1982—had defended German farm interests by ensuring the continued deployment of the complicated switchover mechanism in the new green money regime that came into force with the completion of the Single Market on 1 January 1993. German politics, however, had changed with the unification of the two Germanies, and the farm interests of a united Germany were not the same as those of the old West Germany. In previous coalitions, Kiechle, representing the minority Christian Social Union (CSU) party of Bavaria, had been able to repel the proposals of the other coalition parties, notably the Free Democratic Party (FDP), for a conclusion of the GATT negotiations and reform of the CAP. In the major government reshuffle of January 1993, Kiechle and the CSU lost control of the ministry of agriculture. The new minister of agriculture, Jochen Borchert, was much closer—and more loyal—to the chancellor, Helmut Kohl, and he soon signaled a change in policy. The new minister considered the Blair House Accord to be a "good deal,"

---

3. The Commission claimed to be operating within the formal negotiating mandate given it by the Council under Article 113. If it wished, or needed, to go beyond that mandate it would need to seek the Council's approval, by qualified majority vote. If a member state believed that the Commission had exceeded its mandate, that member state could have challenged the Commission in the European Court. In its memorandum of 26 August 1993 criticizing the Blair House Accord, the French government did claim that "le pré-accord de Blair House ne répond pas aux dispositions du seul mandat, donné à la Commission par le Conseil (en date 6 novembre 1990), sur les questions agricoles" (French Government 1993, 1); but no legal challenge was entered.

claimed that German agriculture would have to become more competitive, and indicated that an early conclusion of the Uruguay Round was in
Germany's interest (*Agra Europe,* 12 February 1993, P3).

It now seemed less likely that a Franco–German coalition would
attempt to block the Blair House Accord and that France was more
isolated. The possibility remained, nonetheless, that France might have
secured a Council vote against the deal or even invoked the Luxembourg Compromise, which would have torpedoed the entire GATT
agreement. As we noted in chapter 3, the Luxembourg Compromise
gives power of veto, despite the provision for qualified majority voting,
provided member states with a combined voting strength of twenty-
three or more refuse to participate in a vote. Whether or not the Luxembourg Compromise still has any operational significance is difficult to
say. It has not been used for a number of years, and some authors have
declared it to be defunct. In the debate in the French Senate over the
Maastricht Treaty before the first Danish referendum, however, the
French government had declared that it retained the "right to impose a
veto in defence of vital national interests, by virtue of the so-called
Luxembourg Compromise" (*Financial Times,* 14 May 1992, p. 3). And,
as we have seen, in November 1992 Prime Minister Beregovoy had
threatened to veto the Blair House Accord.

With French parliamentary elections pending on Sunday, 21 March
1993, and Sunday, 28 March 1993, and with apparently implacable opposition to the Blair House Accord from all parts of the political spectrum
in Paris, the EC Commission and the other member states were anxious
not to aggravate the situation by pressing on with the GATT agreement.[4] However, internal cohesion had to be balanced against the EC's
external credibility. The oilseeds agreement, in particular, was seen as a
symbol of the EC's good intent: if this agreement could quickly be
implemented into EC law, then the EC's trading partners could have
more confidence in the EC's willingness and ability to conclude a meaningful GATT agreement on agriculture.

In February 1993, to take pressure off the farm ministers, the
oilseeds dossier was allocated to the Foreign Affairs Council. But at that
Council's meetings in February and March the matter was not advanced
because of the pending French elections. In April, the French foreign
minister, Alain Juppe, was only a week into his new job, and the matter

---

4. The elections resulted in a right-wing government led by Prime Minister Edouard
Balladur displacing the socialist administration of Pierre Beregovoy and sharing power with
the socialist President François Mitterrand. Presidential elections took place in 1995.

was deferred until May when, again, a vote was not called because the new French government had promised that it was about to make a statement setting out its policy on GATT and oilseeds. Thus, it was not until June 1993 that the Council finally endorsed the U.S.–EC agreement on oilseeds. And this was only after EC farm ministers had agreed to a 27 percent increase in set-aside payments as a sweetener to the French farm lobby (*Agra Europe,* 11 June 1993, E1). France did not invoke the Luxembourg agreement, but she did not have to: France had a de facto power of veto because the Commission and the other member states were unwilling to provoke a "constitutional" crisis within the EC in the aftermath of the acrimonious debates over ratification of the Maastricht Treaty.

Agreement of 15 December 1993

In January 1993 the outgoing U.S. agriculture secretary, Edward Madigan, was still confident that the Blair House Accord provided a satisfactory basis for a multilateral GATT agreement on agriculture, despite French reservations (*Agra Europe,* 15 January 1993, E1). But it soon became apparent that the new Clinton administration—sworn in on 20 January 1993—was less than happy. During Senate nomination hearings, Clinton's nominee as U.S. trade representative, Mickey Kantor, expressed concern about the details of the Blair House Accord (*Financial Times,* 20 January 1993, p. 3). From his comments, it appeared unlikely that the round would be concluded by 1 March 1993, when the president's fast-track negotiating mandate expired. American concerns centered on two issues: namely, the Clinton administration's belief that the EC had secured an extremely generous deal on oilseeds and that the Blair House Accord had not dealt with market access issues. It was on this latter point that Mike Espy, the new agriculture secretary, sought to continue the negotiations (*Agra Europe,* 12 February 1993, P1).

In June 1993, Congress agreed to another extension of the fast-track negotiating mandate. The deal was that President Clinton had to notify Congress by 15 December 1993 of his intention to sign an agreement, which would then have to be submitted to Congress by 16 April 1994. Thus a new, and final, deadline had been set for the conclusion of the negotiations.

With continuing EC opposition to the Blair House Accord, orchestrated by France, the American position changed. Thus, in August 1993, European newspapers were carrying headlines that suggested that the French stance "could sink GATT talks" (*Financial Times,* 31 August 1993, p. 1) unless the Blair House Accord was negotiated in the EC's favor. But the U.S. view was that the deal could not be reopened. At one stage, it appeared as if a Franco–German alliance might reemerge. The breakdown of the narrow bands mechanism of the ERM in August 1993 meant that, in the context of the green money regulations, all EC currencies were floating (see chap. 3 for a more detailed discussion). In particular, this meant that German farmers might experience small reductions in CAP support prices denominated in marks following revaluations of the green mark. In an attempt to secure Council support for a change in the regulations to forestall this possibility, fears were expressed that Germany might support France in its opposition to the Blair House Accord. In particular, following a meeting with the French prime minister, Edouard Balladur, the German chancellor, Helmut Kohl, "hinted strongly that he was prepared to back France in seeking new concessions from the U.S. on farm exports" (*Financial Times,* 27 August 1993, p. 14). These comments were later played down by officials in Bonn.

With the changed composition of the EC Commission at the beginning of the year (see table 6.1), Trade Commissioner Sir Leon Brittan had taken the lead on negotiations on the agriculture dossier and soon found that he and the U.S. trade representative, Mickey Kantor, were able to do business. In the run-up to Blair House, it was the EC farm commissioner and the U.S. agriculture secretary who had led the negotiations, but now agriculture was back in the *trade policy* arena. Nonetheless, the French farm lobby was still in the ascendancy, and France was still determined to renegotiate the Blair House Accord. The dénouement came in the early hours of Tuesday, 21 September 1993, when after twelve hours of negotiations, in a jumbo council of foreign, farm, and trade ministers, Brittan managed to defuse French requests for a reopening of the package (*Financial Times,* 22 September 1993, p. 8). Instead,

TABLE 6.1.   Change of Dramatis Personae, January 1993

United States		EC Commission	
Bush Administration (until 20 January)		1989–92 Commission	
		President:	Jacques Delors
Trade:	Carla Hills	Trade:	Frans Andriessen
Agriculture:	Edward Madigan	Agriculture:[a]	Ray Mac Sharry
Clinton Administration		1993–94 Commission	
		President:	Jacques Delors
Trade:	Mickey Kantor	Trade:	Sir Leon Brittan
Agriculture:	Mike Espy	Agriculture:	René Steichen

[a]Including the agriculture dossier in GATT.

Sir Leon was asked to seek clarification of a number of aspects of the Blair House deal.

However, until the problematic NAFTA agreement had cleared Congress on 17 November 1993, the United States did not feel able to resume high-level discussion. And so the battle of rhetoric continued, with various interests in the EU insisting that the Blair House accord was unacceptable, and the U.S. administration insisting that the Accord could not be renegotiated.[5] It was only once the NAFTA bill was passed that the United States and the EU were free to finalize their accord, take it to Geneva, and present it to the other one hundred-odd participants in the GATT round hoping that they would agree and would conclude the round in the few remaining days before the final deadline.

In the week preceding Wednesday, 15 December 1993, American and European negotiators in Brussels crafted a revision to the Blair House Accord (details of which will be outlined in chap. 7), and the *Financial Times* (9 December 1993, p. 26) was able to report that the French prime minister had claimed a partial victory in the trade talks: "'We have emerged from our isolation without conceding anything essential,' he told cheering deputies from his conservative coalition. . . . Agriculture 'is no longer an obstacle' to agreement." The United States and the EU might have agreed on agriculture, but other dossiers were still proving problematic. Among other outstanding issues, Portugal was threatening to veto the deal unless the United States improved its access offer on textiles (*Financial Times,* 11 December 1993, p. 2). By Sunday evening, 12 December 1993, in Geneva, the package was almost complete, but on Monday the EU and the United States were still "at loggerheads on U.S. access to Europe's film and television markets," with France in particular wanting to protect its culture from an American invasion (*Financial Times,* 14 December 1993, p. 4).

Despite all the last-minute hitches, agreement was finally secured, with hours to spare. When on 15 December 1993 GATT Director-General Peter Sutherland declared in Geneva that the Uruguay Round was concluded after the 117 participants had approved by consensus the 550-page draft treaty, President Bill Clinton had time enough to notify Congress by midnight Washington time that he intended to sign the deal. In Paris, Prime Minister Edouard Balladur had secured a clear vote of confidence in his government and the GATT deal from the French parliament, and the EU's Foreign Affairs Council had unanimously approved

---

5. As of 1 November 1993, with the coming into force of the Maastricht Treaty on European Union, the EC adopted the name "European Union" (EU).

the package. In Paris and Brussels, "a few hundred bedraggled farmers demonstrated . . . , branding the deal a U.S. plot to cut European farming. In Seoul, police arrested hundreds of students protesting against South Korea's acceptance of rice imports" (*Financial Times,* 16 December 1993, p. 1). It is rather ironic that, having been blamed for the impasse in Montreal in December 1988 and for the breakdown in the negotiations in Brussels in December 1990, agriculture should prove to be only one of a number of problem dossiers in Geneva in December 1993 that had threatened to wreck the talks.

The *Financial Times* (16 December 1993, p. 4) was fulsome in its praise for three men:

> Much has happened over the last 12 months to transform the Uruguay Round from a wearisome butt of jokes to a reality that is likely to provide a significant boost to international trade.
>
> The transformation was based on three men: Sir Leon Brittan, appointed on January 1 to succeed the amiable but ineffectual Mr Frans Andriessen as EU trade commissioner; Mr Mickey Kantor, appointed in mid-January as Mr Clinton's U.S. trade representative; and Mr Peter Sutherland, who replaced a battle-weary Mr Arthur Dunkel as director general of the General Agreement on Tariffs and Trade in July.
>
> Their first joint achievement was to set, and then keep in place, yesterday's deadline for completion of Uruguay Round negotiations. GATT's reputation for missing deadlines had almost made it a laughing stock over the seven-year life of the Round.
>
> The chemistry between Sir Leon and Mr Kantor was important. . . . both came to their jobs with reputations as no-nonsense "deal-makers," and their shared commitment to close the Uruguay Round deal was soon apparent.

Marrakesh

Although the trade talks were formally concluded in Geneva on 15 December 1993, a good deal remained to be settled. First, participants had a few weeks to negotiate improved offers before lodging their final tariff and other commitment schedules with the GATT secretariat by 15 February 1994. Second, once all the paperwork had been checked and completed, trade ministers met in Marrakesh on 12–15 April 1994 to sign the agreement. Third, signatories had to ratify the agreement according to their own legislative provisions. It will be recalled from chapter 1 that the 1947 GATT agreement had only ever been applied on a provisional basis because a number of signatories, but most notably the United States, did not ratify the accord. The 1994 GATT agreement provided for the establishment of a World Trade Organization (WTO) recognized in international law, but for this to happen signatories had to formally ratify the agreement for it to come into force in 1995.

For the United States, the legislative procedure was clear. Under the president's fast-track negotiating mandate, Congress could either accept or reject the package in its entirety. But acceptance was not a foregone conclusion. If the timetable for implementing the agreement was to be met, Congress had to agree to the package before beginning its recess on 15 October 1994.

For the EU, the procedure was unclear. It seemed certain that the twelve member states would be members of the new WTO, maximizing the EU's vote in that body, but it was not certain whether the EU as a single entity, or the EU and twelve sovereign member states, would ratify the agreement. In Marrakesh, the agreement was signed by the president-in-office of the Council of Ministers and Sir Leon Brittan, acting as representatives of the EU, and the twelve member states. Although the EU has exclusive competence in matters concerning foreign trade, the agreement also embraced services and other issues where it could be claimed that member states retain jurisdiction. The question was submitted to the European Court for a ruling. At the time *Agra Europe* (5 August 1994, P1) reported that Sir Leon Brittan was of the view that the Marrakesh agreement must be "ratified by all twelve national parliaments to make it legally binding."

Trade and the Environment

At Marrakesh GATT ministers outlined a program of work embracing trade and environmental issues. The Decision of 14 April 1994 on Trade and Environment established a Committee on Trade and Environment that is "to report to the first biennial meeting of the Ministerial Conference after the entry into force" of the new WTO (*GATT Focus,* no. 107, May 1994, 9).

During the Uruguay Round, and particularly in its closing phases, the environmental lobby had grown increasingly hostile to what it saw as a direct clash of interests between the free-trading ethos of GATT and its concerns to develop trade rules designed to protect the environment. In the West, the fear has been that more expensive production practices, inspired by environmental concerns, would be undercut by cheaper imports produced to less exacting environmental standards. Elsewhere the fear has been expressed that import restrictions based upon specious environmental concerns will amount to a new "green" protectionism. In the United States, particular worries were expressed over a GATT panel ruling of September 1991 that had apparently overturned restrictions imposed by the United States on imported canned tuna relating to the use of fishing methods that reduced the likelihood of dolphins becoming entangled and drowned in the nets. However, Palmeter (1993, 66) claims that the ruling centered on the discrimination between U.S. and overseas canners implicit in the rules and not on U.S. attempts to impose rules designed to protect dolphins: "The key is not that the United States imposed standards on the rest of the world; the key is that those standards differed from those it imposed upon itself."

The WTO and the Dispute Settlement Mechanism

The Final Act established the WTO,[6] which will have ultimate responsibility for administering the GATT of 1994, which in turn amends and expands the GATT of 1947,[7] the General Agreement on Trade in Services (GATS), and the Agreement on Trade-Related Aspects of Intellectual Property Rights including Trade in Counterfeit Goods (TRIPS). Members will have to accept the entire package of measures contained in the Final Act: they cannot pick and choose which to apply. The WTO will also administer the various Plurilateral Trade Agreements included in the Final Act, including the International Dairy Agreement and the Arrangement Regarding Bovine Meat. The ultimate decision-making body will be a ministerial conference that will meet every two years but with a General Council to run its affairs in the interim. Decision making will usually be on the basis of consensus, although circumstances will arise in which votes are taken.

A key element in the overall agreement is the Understanding on Rules and Procedures Governing the Settlement of Disputes. This is set out in great detail, in marked contrast to the two articles in GATT 1947 concerned with dispute settlement: "In the Final Act, WTO members have committed themselves not to take unilateral action against perceived violations of the trade rules. Instead, they have pledged to seek recourse in the new dispute-settlement system, and abide by its rules and procedures" (*GATT Focus,* no. 107, May 1994, 12).

---

6. For further details, see *GATT Focus* (no. 107, May 1994) on which this section is based; or Evans and Walsh (1994, chapter 23).

7. GATT 1994 is legally distinct from GATT 1947. Most countries, however, are expected to withdraw from GATT 1947.

The dispute settlements procedure in the GATT 1947 meant that individual parties to a dispute could block panel findings. Under the new arrangements, "there has to be a consensus against the establishment of panels or adoption of panel reports for these decisions not to be made" (*GATT Focus,* no. 107, May 1994, 12). Potentially, this gives the WTO real teeth to police the world trading system: it remains to be seen how it works out in practice.

CHAPTER 7

The GATT Agreement and the CAP

The end product of the Uruguay Round negotiations was *The Final Act Embodying the Results of the Uruguay Round of Multilateral Trade Negotiations,* which was signed by all the contracting parties at Marrakesh in April 1994. This is a complex package deal, incorporating not just an Agreement on Agriculture but also a number of other significant changes to the 1947 GATT, the establishment of the WTO, and a new dispute settlement mechanism, as well as the GATS and TRIPS accords noted in chapter 6. In this chapter we will discuss the Agreement on Agriculture and briefly outline the Agreement on Sanitary and Phytosanitary Measures.

The Content of the Agreement on Agriculture

The Agreement on Agriculture (which excludes fish) is firmly based on Dunkel's Draft Final Act of December 1991 as amended by the Blair House Accord of November 1992 and reflecting concessions made by the United States and the EU in their bilateral discussions in Brussels in December 1993.[1]

There will be three sets of GATT disciplines to constrain farm policies in future years, involving the following:

- internal support;
- import access; and
- export commitments.

---

1. Strictly speaking, the documentation is in three parts: first, there is the legal text entitled Agreement on Agriculture, forming part of the Final Act as signed by all the contracting parties at Marrakesh (GATT 1994a); second, an explanatory document containing the "headline" percentages (20 percent, 36 percent, 21 percent) cited in the following discussion (GATT 1993); and, third, the schedule of commitments actually entered into by each contracting party. Dunkel's Draft Final Act (GATT 1991) had included the numbers now to be found in GATT (1993) rather than in the formal agreement. GATT reports that the commitments tabled by the contracting parties across all aspects of the GATT Agreement amounted to 26,000 pages (*GATT Focus,* no. 107, May 1994, 1).

In the earlier part of the negotiations, the EU had, with some justification, argued that farm support mechanisms should be viewed in their
entirety and that constraints on agricultural protectionism would necessarily have implications for imports and exports. Nonetheless, in the political
bargaining process, a set of complex rules governing all three facets of
farm support emerged. The arrangements, which began to apply in 1995,
set out the new GATT disciplines that are to regulate agricultural trade
after a six-year implementation period, as well as the arrangements that
apply year by year during that implementation period. All developed-
country members of the new WTO have to subject their agricultural
policies to the disciplines outlined below. Developing countries have been
allowed a ten-year transition period and are only expected to achieve two-
thirds of the cuts demanded of developed countries. This "special and
differential treatment" to developing countries was always seen as "an
integral part of the negotiation" (GATT 1994a, Article 15). The least-
developed countries are not obliged to make any cuts.

Furthermore, additional concessions such as the revised arrangements for oilseeds in the EU, stemming from the Blair House Accord,
are built into the country commitments entered into by the signatories.

By late 1994 the EU had not put in place the detailed regulations
that were necessary to implement the agreement in 1995. Exactly why
the European Commission was so tardy in bringing forward proposals
for Council regulations is unclear, but we might speculate that the Commission did not wish to see a reopening in the Council of Ministers of the
debate over the acceptability of the deal before the formal signing cere-
mony in Marrakesh or before ratification of the Uruguay Round by the
U.S. Congress or indeed by the EU itself. The Commission, after all,
had claimed that the GATT deal was compatible with the existing CAP.
One consequence of this delay was that the Farm Council's agenda for
the second half of 1994 looked particularly crowded and controversial.
In addition to modifying CAP regulations to implement tariffication, it
planned to extend the CAP "reform" package to sugar, wine, and fresh
fruit and vegetables and to agree on a new green money system,[2] all
before the end of 1994.

Internal Support

For the CAP, the arrangements dealing with internal support are the
least constraining, but they are not without their own interest. Participat-

---

2. Key provisions of which lapsed on 31 December 1994.

ing countries have had to compute and declare an aggregate measurement of support (AMS) as an annual average for the three-year base period 1986–88. The EU declared a sum of 73.53 billion ecu.[3] This AMS must be reduced by 20 percent over the six-year period of the agreement. However, the contracting parties are allowed to claim "credit . . . in respect of actions undertaken since the year 1986" (GATT 1993, par. 8). Accordingly, the EU has declared the following:

a base period AMS of	73,530,350,000 ecu,
which reduced by 20	
percent would equal	58,824,280,000 ecu,
but with a "credit" of	2,379,900,000 ecu
the bound AMS for the year	
2000 and beyond is	61,204,180,000 ecu per year

(Commission 1994, supporting table 9a).

Exactly how this "credit" has been calculated, or why it should be allowed, is unclear.

The Draft Final Act envisaged the calculation being undertaken on a product-specific basis. The Blair House Accord very much weakened this provision by providing for an AMS for the farm sector as a whole. Note that the original formulation would not necessarily have reduced raw material prices for the food industry or consumers because a reduction in the AMS could be achieved by a tightening of quotas, thereby curbing the volume produced. However, a rise in world market prices does not reduce the AMS because the calculations for all subsequent years will take 1986–88 world prices as the benchmark.

There is a *de minimus* provision as well. In determining the AMS, countries are not required to include in the calculations, or subsequently reduce, any product-specific support "where such support does not exceed 5 per cent by value of that Member's total value of production of a basic product during the relevant year." For developing countries, the *de minimus* percentage is 10 percent (GATT 1994a, Article 6).

A 20 percent cut in the AMS sounds harsh, but over six years its effects would be muted. If EU farm support prices are double world market prices (not an unreasonable assumption), then a 20 percent cut in support implies a 10 percent cut in price. Given productivity gains of 1

---

3. Supporting Table 9a, Commission (1994). All of the EU's GATT commitments are expressed in terms of commercial ecu. See chapter 3 for further explanation.

to 2 percent per year in the farm sector and little growth in demand, a price decline of this magnitude over six years is unremarkable.[4]

It was always intended that any aid paid to the farm sector that had no discernible impact on production ("decoupled" payments, to use the GATT jargon) would be exempt from these GATT disciplines. Thus, various "general" government services such as research funding, extension and advisory work, and inspection services would be exempt, as would expenditures on "public stockholding for food security purposes" and "domestic food aid." In addition, exempt "direct payments to producers" include "decoupled income support," "government financial participation in income insurance and income safety-net programs," "payments for relief from natural disasters," "structural adjustment assistance," and payments under "environmental programs" or "regional assistance programs"—but all subject to the constraint that "the support in question shall not have the effect of providing price support to producers." All these payments have to be provided through "a publicly-funded government program (including government revenue foregone) not involving transfers from consumers" (GATT 1994a, various items from Annex 2).

At Blair House, the Americans and Europeans stretched this concept of "decoupled income support" payments to include both the deficiency payments of the United States and the EU's area compensation and headage payments introduced under the Mac Sharry reforms. The agreement (GATT 1994a, Article 6.5[a]) does not list these payments as exempt; it merely sets out the following requirements:

Direct payments under production-limiting programmes shall not be subject to the commitment to reduce domestic support if:

(i)   such payments are based on fixed area and yields; or
(ii) such payments are made on 85 percent or less of the base level of production; or
(iii) livestock payments are made on a fixed number of head.

Article 13.2 of the agreement does state that these payments are only exempt provided that the level of support to a particular commodity does not exceed that granted in the 1992 marketing year. Nonetheless, these conditions are very much less onerous than those required elsewhere in the text on "decoupled income support" (GATT 1994a,

---

4. A note of caution: inflation would erode the real value of the aggregate measurement of support over time, and the outlook for EU inflation in terms of ecu is uncertain.

Annex 2, Article 6) and are clearly designed to include the EU's schemes. Inclusion of the EU's area compensation and headage payments in the exempt category has a magical impact on the EU's AMS calculation because the AMS base period includes the old system of supporting cereal growers and livestock producers, while future area compensation and headage payments will be excluded from the calculations. The "credit" earned in the cereals and livestock sectors can be used to avoid cutbacks elsewhere. Furthermore, having established the precedent that the Mac Sharry area compensation and headage payments are exempt from GATT disciplines, the temptation must be to extend the principle into other sectors—subject only to the budgetary constraint and newspaper reports about large "wealthy" farmers receiving annual area compensation payments of £250 per hectare without means testing.

In determining the AMS, clearly any "nonexempt direct payments" are included in the calculations, as is the value of any "market price support." The latter is "calculated using the gap between a fixed external reference price (based on the years 1986 to 1988) and the applied administered price" multiplied by the quantity of eligible production but excluding any "Budgetary payments made to maintain this gap, such as buying-in or storage costs" (GATT 1994a, Annex 3). Thus, if the EU allows its intervention stocks to soar in future years with exports constrained by the GATT disciplines, the resulting budgetary costs will not count in the AMS.

Import Access

There are four basic elements to the agreement:

- "tariffication": that is, all existing border measures are to be converted into tariffs;
- all tariffs are to be reduced by a minimum of 15 percent, and on average by 36 percent, in six equal steps over the transition period;
- a minimum access clause: countries are to make arrangements to allow imports to capture a minimum of 3 percent of the home market in the first year, rising to 5 percent in year six; and
- a special safeguards clause: countries may apply additional duties if they face an import surge or if the landed price of a particular consignment falls below a trigger price equal to the average import price recorded in the period 1986–88. In practice, this means that tariffication results in a less than pure version of an import tariff.

The tariff equivalents to be calculated as a result of tariffication are essentially based on the difference between "a representative wholesale price ruling in the domestic market or an estimate of that price where adequate data is not available" and the "actual average cif unit values for the importing country" or, when the CIF value is not available or appropriate, "appropriate average cif values of a near country" or an estimate based on average FOB values from an appropriate major exporting country with the addition of "an estimate of insurance, freight and other relevant costs" (GATT 1994a, attachment to Annex 5).[5] After the Blair House discussions, the Commission made clear that it intended to take a product's "intervention price on the Community market, increased by 10 percent and by the monthly increments" as indicative of the wholesale price (Commission 1992). Whether "a representative market price" and the intervention price plus 10 percent are equivalent valuations is a moot point, but, given the size of the tariff equivalents established, any difference is trivial. It should be noted nonetheless that for many products the intervention price plus 10 percent is less than the corresponding threshold price for the product, indicating that tariffication in itself involved a lowering of import protection by the EU. For example, for cereals this reduces the tariff equivalent by about 40 ecu per metric ton.

Details of some of the calculations were given in an Agra Europe report based on Commission documentation (Agra Europe 1993). These are reproduced in table 7.1 to give an indication of the levels of import protection actually afforded by the CAP. As can be seen, for all the products listed the import tariffs are substantial.

For products for which variable import levies used to apply, *specific* tariff equivalents in ecu have been fixed. Thus, as previewed in table 7.1, the definitive "base rate of duty" declared by the EU for common wheat was 149 ecu per metric ton. Where ad valorem duties applied, they continue as ad valorem duties in the new system. For beef, both specific and ad valorem duties continue to apply. For certain fresh fruits and vegetables, for specified periods of the year the EU intends to continue with a minimum entry price system in addition to ad valorem duties. For agricultural and food products, excluding fish, the EU's tariff schedule as declared to GATT covers 108 pages plus annexes.

---

5. CIF is an acronym regularly used in international trade, and means "cost, insurance, freight": thus, it is equivalent to the price of a product at a country's border before payment of any import taxes. Similarly, FOB means "free on board" and is equivalent to the price of a product loaded on board ship in the country of export.

For most products, the EU will reduce its tariffs by 36 percent over the six-year implementation period, although for some items the reduction will be 20 percent. For products of little interest to the EU, or where trading partners have pressed for additional concessions, the reduction is sometimes 50 percent and very occasionally 100 percent. In no case has the EU offered only the minimum 15 percent cut nor is there evidence to suggest that the Commission engaged in strategic balancing of tariff lines to achieve the required arithmetical average of 36 percent by making major cuts for unimportant items and minor ones for more "sensitive" products. Sugar, most fruit and vegetables, and some milk products faced a tariff cut of 20 percent. But most mainstream CAP products will see tariffs falling by 36 percent. Thus, for common wheat, the base rate of duty of 149 ecu per metric ton will be cut to the following:

140 ecu per metric ton on 1 July 1995
131 ecu per metric ton on 1 July 1996
122 ecu per metric ton on 1 July 1997
113 ecu per metric ton on 1 July 1998
104 ecu per metric ton on 1 July 1999
95 ecu per metric ton on 1 July 2000 ("new bound rate of duty").

However, the import duty on cereals is subject to additional constraints, as outlined below.

TABLE 7.1.   Determination of the EU's Tariff Equivalents for Selected Products (commercial ecu per metric ton)

	Internal Price 1986–88	External Price 1986–88	Tariff Equivalent	Average Import Levy 1986–88
Common wheat	241	93	149	195.7
Barley	236	85	145	188.2
Maize	241	96	147	182.4
Whole milled rice	914	235	680	752.0
White sugar	719	196	524	542.0
Frozen beef carcases[a]	3,436	1,423	2,013	
Skim milk powder	2,170	685	1,485	
Butter	3,905	943	2,962	

Source: Columns 1 to 3, Agra Europe (1993, 8); column 4, European Community (1990, Annex II).
[a] plus a 20 percent ad valorem duty.

To implement the minimum access clause, the EU has decided to open tariff quotas at 32 percent of the basic tariff.[6] The tariff quotas are set equal to any shortfall in average annual imports over the 1986–88 base period, compared to 3 percent (rising to 5 percent) of home consumption in that base period. Thus, if annual consumption in the base period for a relevant product amounted to 100 thousand metric tons and imports to 2 thousand metric tons, then a tariff quota for 1 thousand metric tons would be opened for the first year of the transition period, rising to 3 thousand metric tons at the end of the period. This provision mainly affects pig and poultry meat and cheeses. Where special import arrangements already apply, as with New Zealand butter into the EU, the 1986–88 access arrangements must be sustained. Thus, the EU is required to revert to its 1986–88 level of New Zealand butter imports of 76,667 metric tons per year compared with the 1993 tariff quota of 51,830 metric tons.

Tariff quotas, particularly for such small quantities of product in relation to the size of the domestic market, would usually result in the generation of quota rents. Whoever controls the tariff quota, whether in the form of an export license issued by the beneficiary country or an import license issued by the importing state, would normally be able to expropriate the unpaid duty for their own use. Depending on how the quota is allocated, there is no guarantee that the exporting state will gain an increase in revenues. If the authorities in the importing state auction off the quotas or allocate licenses to their own traders, for example, the benefits will be retained at home. We suspect that the EU will strive to ensure that the financial benefit to be gained from the new tariff quotas to be introduced as a consequence of the Uruguay Round will be reaped by EU rather than overseas traders. Exporters will naturally resist this. In discussing the allocation of "entitlement for access to quota restricted markets" as a result of the new access arrangements in the Uruguay Round, Australia's Industry Commission has suggested that these rights should be auctioned off, with the revenues accruing to the Australian government (Industry Commission 1994, 75). It also reports that in 1993 quota rents on access to the restricted U.S. meat market amounted to 1.4 Australian dollars per kilogram and for the EU sheep and goat meat market rents were 1 Australian dollar per kilogram (Industry Commission 1994, 71).

There are a number of supplementary arrangements, with the EU opening tariff quotas for maize and sorghum and offering additional

---

6. This figure of 32 percent is first reported in Commission (1992, 2).

reductions in tariffs on a number of products, notably fruit and vegetables, of export interest to the United States. The Japanese and South Korean rice markets have a special status: for example, Japan will only apply tariffication for rice after six years but will open up its market to allow imports to capture 8 percent of consumption by the year 2000.

Special Safeguard Provisions

As we noted in the previous section, tariffication results in a less than pure version of an import tariff in that the agreement contains special safeguard provisions that allow for the application of additional duties under two market circumstances:

- when the country experiences a surge of imports, or
- when the CIF import price of the shipment concerned falls below a trigger price equal to the average import price recorded in the period 1986–88 (GATT 1994a, Article 5).

However, this second clause can only be invoked if tariffication has been introduced as a part of the Uruguay Round agreement.

In the first instance, an additional import duty not exceeding "one third of the level of the ordinary customs duty" can be applied to all imports of the product for the remainder of the calendar year. In the second case, an additional import duty determined on a sliding scale can be charged on the particular shipment concerned. These additional duties cannot be simultaneously applied, and detailed rules are laid down.

The formula for determining the additional duty to apply when the import price falls below the 1986–88 reference level is complex: the first 10 percent decline below the 1986–88 reference level is ignored; for the next tranche of 30 percent (i.e., between 10 and 40 percent below the 1986–88 reference level), a 30 percent additional duty can be applied, and so on. The effect is illustrated in figure 7.1 where the price decline below the 1986–88 reference level is measured on the horizontal axis, and the 45 degree line maps out the additional import levy that would be applied under the existing variable import levy mechanism. As figure 7.1 shows, the agreement falls far short of this and indeed is much less than the EC had suggested in its November 1990 GATT offer (European Community 1990).

In its GATT offer of November 1990, the EC had asked for a "corrective factor" that would "offset all currency fluctuations," whereas the agreement does not distinguish between the causes of a fall in CIF import prices expressed in the importing state's currency (ecu for the EU).

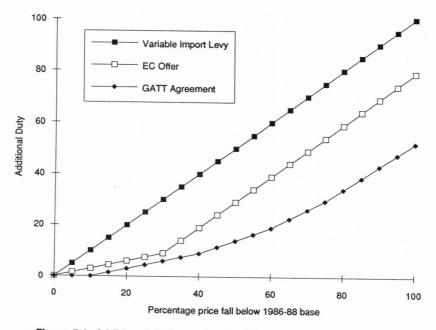

Figure 7.1. Additional duties under the GATT agreement, consequent upon an import price falling below the 1986–88 reference level

Nonetheless, although the agreement does fall far short of the EU's request, it does amount to a partial reinvention of the variable import levies when world prices fall below their 1986–88 levels.

The way the system works depends critically upon the 1986–88 "reference prices" that trigger the mechanism. The GATT agreement declares that the reference price should "be publicly specified and available to the extent necessary to allow other Members to assess the additional duty that may be levied" (GATT 1994a, footnote to Article 5). If this were not the case, the new system would be even less transparent than the existing regime of variable import levies. Nonetheless, sufficient scope exists for the rules to be exploited.

Three World Market Prices

It might be concluded that the 1986–88 reference price to be used as the trigger for the safeguard clause would be the same as the 1986–88 world price that the EU had computed for the tariffication exercise. However, this is not the case. In fact, the agreement allows countries to deploy three different world market prices in determining their commitments.

For the determination of the tariff equivalent, the external price should, "in general," be the "actual average cif unit values for the importing country," but where these are not "available or appropriate" countries are allowed to use "appropriate average cif values of a near country" or an estimate based on "average fob unit values of (an) appropriate major exporter(s) adjusted by adding an estimate of insurance, freight and other relevant costs to the importing country" (GATT 1994a, attachment to Annex 5). For wheat, the EU has used an FOB price in a major exporter, and it has thus reported an external price for 1986–88 of 93 commercial ecu per metric ton (as shown in table 7.1).

For the AMS calculation, "The fixed external reference price shall be based on the years 1986 to 1988 and shall generally be the average fob unit value for the product concerned in a net exporting country and the average cif unit value for the product concerned in a net importing country in the base period. The fixed reference price may be adjusted for quality differences as necessary" (GATT 1993, Annex 5, par. 9). As a net exporter of wheat, the EU has used an EC FOB price for 1986–88 of 86.5 commercial ecu per metric ton (Home-Grown Cereals Authority 1994, 6).

For the special safeguards provisions, the relevant 1986–88 reference price should in general be "the average cif unit value of the product concerned" (GATT 1994a, footnote 2 to Article 5). As an importer of high-quality bread-making wheats, the EU's average CIF value of wheat imports in the reference period has been declared to be 148 commercial ecu per metric ton (Home-Grown Cereals Authority 1994, 4). This is clearly much higher than either the world price used for the AMS or tariffication computations.

Similarly, the authors understand that for sugar the EU has argued that the appropriate reference price to adopt under the special safeguards provisions is that of the imports of ACP sugar into the EU at that time and not the very much lower world market price. Under the Lomé Convention, certain ACP (African, Carribbean, and Pacific) states are able to supply specified quantities of sugar to the Community at the EU price.

Maximum Import Duty for Cereals

At Blair House, the EC acceded to a U.S. request that it should impose further constraints on the level of the import duty applicable for cereals and rice. In particular, for cereals the EU has agreed "to apply a duty at a level and in a manner so that the duty-paid import price for such cereals will not be greater than the effective intervention price (or in the event of a modification of the current system, the effective support price) increased by 55 percent" (Commission 1994). It is easy to understand the rationale for this move: the Mac Sharry reforms had introduced a July threshold price of 155 green ecu per metric ton, some 55 percent higher than the intervention price.[7] With a move to a fixed tariff based on the pre-Mac Sharry reform prices, the Americans were clearly fearful that the new system would systematically result in landed prices in excess of 155 green ecu per metric ton.

Tangermann and Josling (1994) claimed that the 55 percent test must apply to each shipment, and from this they concluded that the EU was likely to determine the applicable import duty on an individual consignment basis. Thus, as with the additional duties to be determined under the safeguard clause, when import prices fall the import duty on cereal imports would be a voluntary tax if traders could produce invoices acceptable to the customs officers showing that the landed price is sufficiently high to avoid payment of duty. This is hardly the outcome the EU could have intended, but it is a potential risk that results from asking traders to respect minimum import prices.[8] Under the old reference price system for fresh fruit and vegetables, which is displaced by the Uruguay Round Agreement, the EU sought in part to overcome this problem by monitoring the price at which products were subsequently sold within the EU. Traders were allowed to minimize the import tax paid by charging the EC a higher price for imports, but they were not permitted to pass on to their customers the benefit of paying a reduced import tax either in the form of lower prices or in under-the-counter discounts. There is a clear advantage for EU-based companies in setting up Swiss subsidiaries to ensure that minimal rates of import tax are charged and that the benefits are not repatriated to the EU. Even if arrangements are put into force that ensure that traders cannot choose the CIF price they declare but instead are forced to declare an import price representative of world market conditions, there is a danger that the cap imposed upon import taxes on cereals and rice by the U.S.–EU deal will result in all quality grades being imported at the same price. This would be the outcome if, for example, the EU applied import

---

7. Equivalent to 184.7 commercial ecu per metric ton, after allowing for green currency changes.

8. On the EU's earlier minimum import price regimes, see Ritson and Swinbank (1986) and Harris and Swinbank (1991).

tariffs so that the effective tariff-paid price for high-quality bread-making wheat and that for low-quality feed wheat was 184.7 commercial ecu per metric ton.

Export Commitments

The Uruguay Round Agreement imposes two constraints on the use of export subsidies. These are that annual expenditure on export refunds has to be reduced by 36 percent compared with a 1986–90 base and that the volume of subsidized exports (excluding processed foods) has to be reduced by 21 percent (the Dunkel text had proposed 24 percent) compared with the 1986–90 base.[9] Originally, it had been proposed that these reductions be "front-loaded." That is, in year one they implied a 6 percent and 3.5 percent reduction from the base in expenditure and volume, respectively. It was these proposed export constraints that caused the greatest controversy in the EU and particularly in France. The "problem" is illustrated in figure 7.2.[10]

In each graph of figure 7.2, the first column indicates the 1986–90 base and the fourth column the maximum annual volume of exports permitted by the end of the implementation period. The second column indicates the actual level of exports in 1991–92. Clearly, for some products recent exports significantly exceed the 1986–90 annual average that forms the reference level. Thus, under the original proposal, cheese exports would have been cut back very significantly in year one of the agreement. A contrary situation applies for butter and skim milk powder (SMP). Here, recent export levels are below those established for the 1986–90 base, and indeed the EU is entitled to increase its exports of butter and SMP. These questions were of more than passing interest to the EU's dairy industry and were the focus of an EU–U.S. compromise in the first week of December 1993. The agreement retains the year six target as a 21 percent reduction on the 1986–90 base, but where annual average 1991–92 exports (1986–92 for beef) exceed the 1986–90 base there are to be equal annual reductions from the 1991–92 export level with some transfer possible between years. This is reflected in column 3, which shows the maximum volume of exports in 1995. For some products, such as butter, this is scaled back from the 1986–90 base whereas for others, such as cheese, the reference level is 1991–92. These arrange-

---

9. Genuine food aid shipments are exempt.

10. Data supplied by the Commission of the European Communities and reported in Agra Europe (1993).

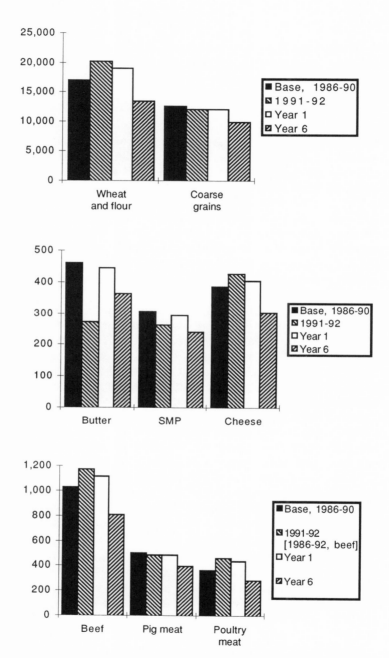

Figure 7.2. The "front-loading" problem for cereals, dairy, and meat: EU exports (thousand metric tons per year)

ments will allow the EU (and the United States) to subsidize the export of additional quantities of surplus products over the implementation period, relieving the pressure on intervention stocks but dampening prospects for an early increase in world market prices. For example, for the EU this amounts to another 8.1 million metric tons of wheat and wheat flour, a 10 percent increase in the quantities that could have been exported over the five-year period 1995–99 if the originally proposed profile had been adopted.

Processed Foods

The Agreement on Agriculture, in imposing similar constraints on the use of export subsidies for processed food products, in effect legitimized the use of such subsidies where they had previously been challenged. Recall from chapter 3 that an unsettled issue in the simmering trade disputes between the EC and its partners in the 1980s had been the EC's refusal to accept a GATT panel ruling of 1983 that export subsidies paid on pasta could not be considered a legitimate use of the GATT-sanctioned ability to pay export subsidies on primary products, notwith-standing the incorporation of a primary product—durum wheat—in pasta. Article 12 of the agreement (GATT 1994a) limits the subsidy paid "on an incorporated agricultural primary product" to that which "would be payable on exports of the primary product as such." Furthermore, the export subsidy expenditure discipline treats expenditure on incorpo-rated products as a separate item, although there is no corresponding constraint on the volume of subsidized exports.

Despite this favorable outcome, the EU's food industries fear that their interests will be adversely affected by the agreement, and the industry lobby understandably points out this concern. Harris (1994, 7–8) notes that annual expenditure on export refunds on non-Annex II goods had risen to 702 million ecu in 1991 and 1992 compared to 572.5 million ecu in the 1986–90 base period. Thus, the cutback in spending over the six-year implementation period will be in the order of 48 per-cent. He goes on to quote a letter of July 1994 sent by the European food industry association (CIAA) to Commissioners Brittan, Steichen, and Bangemann, which, in part, said

> Exports of processed foodstuffs should be given priority over ex-ports of unprocessed agricultural commodities because, by adding value these benefit the whole of the Union economy to a greater degree, without reducing the overall volume of exports of raw mate-rials. (Harris 1994, 9)

Of course, many of the export constraints on agricultural products in fact apply to products of the first-stage processing industries (white sugar, milk powder, cheese, etc.) and thus imply plant closures or a reduction in the utilization of production capacity. These businesses, as well as farmers and the input supply industries, face adjustment problems as agricultural reform progresses.

Bananas

Bananas proved to be a problem commodity for the EU. As a result of the original Treaty of Rome, Germany had been allowed to import bananas duty-free from any source and had obtained its supplies principally from Latin America, although the EU's normal import tariff stood at 20 percent, reduced to zero for ACP suppliers. However, as a result of policies pursued before creation of the EC, a number of member states continued to regulate their domestic markets to give preferential access to particular sources. For the United Kingdom, these were former colonies in the Caribbean; for France, French producers in the DOM (*départements d'outre-mer*); and for Spain and Portugal, suppliers in the Canary Islands and Madeira. With a single market due to come into being on 1 January 1993 these separate market regimes were unsustainable. However, it was not easy to devise a single EU import regime that would not increase the price of bananas to German consumers, that would allow EU and ACP producers to retain a privileged position in the European market, and that would not outrage the Latin American suppliers who hoped to expand their trade with the EU as a result of a more liberal import regime. Any attempt to raise import barriers for bananas entering the German market would cause difficulties in the GATT, and would be contrary to the intent of the Uruguay Round.

The EC's Council of Ministers finally put together a package deal in December 1992, although this faced legal challenges from Germany in the European Court, and German consumers were outraged over the impact of a CAP regime on this particular food price. Community producers benefit from protected market prices and a deficiency payments scheme. Traditional quantities of ACP fruit are admitted to the EU free of import duty. Within a tariff quota of 2 million metric tons, later increased to 2.118 million metric tons, other suppliers—often referred to as dollar-zone producers—pay an import duty of 100 ecu per metric ton, subsequently reduced to 75 ecu per metric ton. Imports in excess of this tariff quota would pay a prohibitive tariff of 850 ecu per metric ton. In order to ensure that ACP supplies can compete, some 30 percent of the tariff quota of dollar-zone bananas is allocated to the ACP shippers on the basis of their import volumes of ACP fruit. This is clearly de-signed as a cross subsidy—the profits from shipping dollar-zone bananas offset the additional costs of ACP supplies—but in reality the beneficia-ries may well be the shippers.

A group of Latin American countries challenged the new regime in the GATT, but Costa Rica, Colombia, Nicaragua, and Venezuela with-drew their complaint as a result of the increased tariff quota and the reduction of the within tariff-quota tariff rate, conceded by the EU. This framework agreement between the EU and the four Latin American countries, valid until 31 December 2002, has prompted a group of U.S.-based banana shippers, including Chiquita Brands, to ask the U.S. trade representative to investigate (under the Super 301 procedures) whether the EU's import licensing system favors EU-based companies at the expense of those domiciled in the United States (*Financial Times,* 8 September 1994, p. 36; and 23 September 1994, p. 34). The EU had believed that the Uruguay Round settlement meant that the United States should not adopt unilateral actions in international trade disputes but should instead make use of the disputes settlement procedures agreed upon. Bananas may well prove to be the focus of the first of the post-Uruguay Round trade wars between the United States and the EU.

Oilseeds

For many years, the United States had complained that the subsidy systems introduced by the EC in the 1970s to encourage the production of oilseeds negated GATT bindings conceded by the EC in the 1960s. Two GATT arbitration panels upheld this complaint. The EC had hoped to resolve this issue by "rebalancing" agricultural protection in the con-text of the GATT negotiations, but the Americans were unwilling to countenance rebalancing in any meaningful way. Faced with this im-passe, the United States had threatened to introduce punitive import tariffs in December 1992 on a range of agricultural and food products, carefully selected to target the EC in general and France in particular. The U.S.–EC accord of November 1992 diffused this dispute. At the time, this was reported as a separate bilateral deal clearly connected with, but not formally part of, the Uruguay Round. The EU has nonethe-less incorporated the Memorandum of Understanding on Oilseeds into its formal schedule of commitments of bindings on domestic support (Commission 1994).

The EU will be able to continue subsidizing oilseed production by payment of area subsidies. However, if the total subsidized area planted exceeds a predetermined level, the area subsidy payable will be reduced

proportionally. This outcome is much less cumbersome than what had
been feared. At one stage it had seemed that the Americans might insist
upon limits to the area planted. Furthermore, the agreement excludes
"unsubsidized" oilseeds grown on set-aside land for "industrial purposes," provided the by-products do not exceed the equivalent of 1
million metric tons of soybean meal equivalents.[11] This would seem to
open up a new business opportunity for European farmers because under the May 1992 reforms they receive area payments on set-aside land
even if used for "industrial" crops. Differential taxation could allow the
absorption of the vegetable oil as a diesel replacement for motor fuel,
making 1 million metric tons of meal available for animal feed. When
American soybeans are imported into the EU for crushing, it is for the
meal that they are sought, with much of the soybean oil reexported onto
world markets. Thus, the EU appears to have achieved a remarkably
generous deal for Europe's farmers.

Rebalancing

Although rebalancing had been an important component of the EC's
negotiating strategy, it was in effect abandoned at Blair House and failed
to be reflected in the final agreement of December 1993. At Blair
House, a bilateral agreement between the United States and the EC was
nonetheless agreed upon. According to the Commission (1992, 4), this
agreement reads as follows:

> If EC imports of nongrain feed ingredients increase to a level, in
> comparison with the level of imports 1986–1990, which undermines
> the implementation of CAP reform, the parties agree to consult
> with a view to finding a mutually acceptable solution.

It remains to be seen what action, if any, the United States would
contemplate if the EU wished to implement this understanding.

Peace Clause

For the EU, an important outcome of the negotiations was the acceptance of the CAP by its trading partners and an assurance that the

---

11. Equivalent to about 2.3 million metric tons of rapeseed before crushing. When
oilseeds are crushed they yield oil and a protein rich residue (meal) much valued for
animal feed.

support provisions painfully negotiated during the Uruguay Round would not be challenged as soon as the ink was dry on the agreement. Dunkel's Draft Final Act had, however, only made the limited promise that

> Where reduction commitments on domestic support and export subsidies are being applied in conformity with the terms of this Agreement, the presumption will be that they do not cause serious prejudice in the sense of Article XVI:1 of the General Agreement. (GATT 1991)

At Blair House, the United States and the EC totally rewrote and expanded this statement, making clear that domestic support measures and export subsidies that conform fully with the Agreement on Agriculture will, "during the implementation period," be exempt from challenge (GATT 1994a, Article 13). Somewhat confusingly, in the context of Article 13 the "implementation period" is of nine years' duration, commencing in 1995, rather than the six-year implementation period defined elsewhere in the agreement (GATT 1994a, Article 1). In its review of the Blair House Accord, the Commission stated:

> This text, in substance, lays down that:
>
> - internal support measures are exempt from actions undertaken under Article 16 of the GATT, as well as actions for nullification and impairment. . . .
> - export subsidies are exempt from claims under Article 16 of the GATT. This eliminates the risk of panels.
>
> This obviously only applies where the provisions of the agreement on agriculture are respected. Countervailing duty actions remain possible but are subject to conditions which make them unlikely. (Commission 1992, 4)

In conclusion, the Commission remarked that "the CAP is 'safe' under the legal rules of GATT because of the adoption of the 'peace clause,'" and "The CAP is now compatible with GATT and recognised as such" (Commission 1992, 10).

Despite this confident assertion, the peace clause only applies during a nine-year implementation period and thus lapses in the year 2004. Nonetheless, by then a further round of GATT negotiations concentrating on agricultural protection is scheduled. In the words of the Agreement on Agriculture (GATT 1994a, Article 20):

> Recognizing that the long-term objective of substantial progressive reductions in support and protection resulting in fundamental reform is an ongoing process, Members agree that negotiations for continuing the process will be initiated one year before the end of the implementation period [of six years in this instance], taking into account:

> — the experience to that date from implementing the reduction commitments;
> — the effects of the reduction commitments on world trade in agriculture;
> — nontrade concerns, special and differential treatment to developing country Members, and the objective to establish a fair and market-oriented agricultural trading system, and the other objectives and concerns mentioned in the preamble to this Agreement; and
> — what further commitments are necessary to achieve the above mentioned long-term objectives.

Thus, while it must be conceded that the actual outcome of the Uruguay Round falls far short of President Reagan's initial suggestion of a "zero option," the process of reform has begun. The EU will again find itself isolated and the CAP challenged by its trading partners within the new WTO if it fails to offer further "substantial progressive reductions in support and protection" in the negotiations to come.

Sanitary and Phytosanitary Measures

The *Agreement on the Application of Sanitary and Phytosanitary Measures* is an integral part of the overall Final Act (GATT 1994b). Sanitary and phytosanitary measures are defined as any measure applied in the following ways:

> — to protect animal or plant life or health within the territory of the Member from risks arising from the entry, establishment or spread of pests, diseases, disease-carrying organisms or disease-causing organisms;
> — to protect human or animal life or health within the territory of the

Member from risks arising from additives, contaminants, toxins or disease-causing organisms in foods, beverages or feedstuffs;
— to protect human life or health within the territory of the Member from risks arising from diseases carried by animals, plants or products thereof, or from the entry, establishment or spread of pests; or
— to prevent or limit other damage within the territory of the Member from the entry, establishment or spread of pests. (GATT 1994b, 8)

Over many years, sovereign states have developed a range of such measures reflecting their understanding of scientific knowledge, the level of income and preferences of their populations, and the particular geographic, climatic, and cultural influences and eating preferences faced. Despite the efforts of bodies such as the Codex Alimentarius Commission to achieve some degree of harmonization in the measures applied, considerable differences persist. Inevitably, these act as potent nontariff barriers to trade and frequently give rise to the suspicion that they are applied with a protectionist intent. Petrey and Johnson (1993, 434) noted that the objective in the Uruguay Round negotiations was "to establish a common set of rules and disciplines to guide the adoption, development and enforcement of such measures."

The lobbyists—also reflecting the ideas expressed in the related concerns over environmental protection—developed two apparently opposing theses. Producers with an export interest expressed concern that importing countries might put in place specious sanitary and phytosanitary measures unsupported by scientific evidence, which were simply designed to protect their domestic producers from cheaper imports. The EC's ban on the use of growth hormones in beef production was cited as a case in point. On the other hand, consumer and producer lobbyists in countries with "high" standards were fearful of any attempts to impose legislation at an international level as this would imply a lowering of their standards.

It should be emphasized that the GATT agreement does not impose sanitary and phytosanitary standards at an international level, but it does try to ensure that unwarranted trade barriers that cannot be backed up by scientific evidence are eliminated. It states quite clearly that "no Member should be prevented from adopting or enforcing measures necessary to protect human, animal or plant life or health, subject to the requirement that they are not applied in a manner which would constitute a means of arbitrary or unjustifiable discrimination between Members where the same conditions prevail or a disguised restriction on international trade" (GATT 1994b, 1).

Sanitary or phytosanitary measures based on international standards are deemed to be compatible with GATT. Members are required to recognize the equivalence of trading partners' measures, and the agreement emphasizes the need for transparency in the regulations: the measures should be "published promptly," and both local producers and importers should have equal access to information and to testing and control procedures.

If disputes between trading partners do arise then the new Understanding on Rules and Procedures Governing the Settlement of Disputes is to apply. Whether or not these new arrangements will have any practical effect on the outcome of trade disputes centering on sanitary and phytosanitary measures is difficult to say. At one extreme, some commentators suggest that nothing will change, whereas others seem to believe that the new WTO has been given significant powers to police the application of trade-distorting sanitary and phytosanitary measures. Harris (1994b, 3) suggests that the EU's ban on the import of beef from cattle dosed with growth hormones, which he claims was "adopted as a result of consumer pressure in the European Parliament" with "no scientific basis . . . for the measure," may well prove to be "a test case for the Uruguay Round Sanitary and Phytosanitary Agreement."

Impact on the CAP

From the foregoing discussion it will be evident that the constraints of the Uruguay Round Agreement on internal support imply no further adjustment to the CAP. Tariffication, the phased reduction in tariffs, and the minimum access clause arrangements do imply some technical changes to the import regimes, but, for cereals at least, the variable levy mechanism will in practice remain, and community preference will be preserved. As outlined earlier in this chapter, there is a danger that, depending on the precise implementation arrangements, traders will avoid paying any import tax on cereals and rice or supplementary levies under the safeguards clause by ensuring on a shipment basis that their declared import price does not trigger import charges.

It would appear that the main constraints relate to export controls, where there are both technical and policy issues to be addressed. If export volumes and/or expenditure on export subsidies have to be controlled, the Commission could pursue one or more of the following strategies:

- reduce the level of export refunds and hence the financial incentive to export;
- extend the system of tendering by traders to determine export refund levels and the volume of exports sanctioned by export licenses; and
- ration the allocation of export licenses, either on a first-come, first-served basis or according to past export performance.

If the export controls prove to be a binding constraint, intervention stocks are likely to rise unless domestic production is constrained by tighter quotas or more set-aside, or production is discouraged and consumption expanded as a result of lower CAP prices. Much debate has centered on cereals.

After Blair House, the Commission claimed that, for cereals, the Blair House Agreement was in accordance with the Mac Sharry reforms and that no further adjustment to the cereals regime would be needed. It suggested, for example, that as a result of the lower cereal prices introduced by the Mac Sharry reforms, livestock consumption of cereals would increase both because of a displacement of cereal substitutes and because of an increase in consumption of pig and poultry meats. The overall increase in cereal utilization was put at 12 million metric tons per year by 1999–2000 (Commission 1992, 8).

Its "worst-case scenario" was that yields would increase by 1 percent per year following the Mac Sharry reforms. With an assumed base of 5 metric tons per hectare, rather more than the actual yield recently achieved, production would reach 176.8 million metric tons in 1999–2000, giving rise to an exportable surplus in excess of the GATT constraints of 2 million metric tons per year at the end of the implementation period (Commission 1992, 8).[12] The final agreement of December 1993 resulted in slightly less onerous export constraints during the implementation period, and consequently—if the Commission's 1992 predictions are accurate—its worst-case scenario would not result in an accumulation of stocks, or the need for additional set-aside, during the implementation period. In its document, however, the Commission did not address the longer-term balance sheet beyond the six-year implementation period.

Moreover, the Commission's view was that a 1 percent annual increase in yields, with a starting point of 5 metric tons per hectare, was

---

12. In fact, the Uruguay Agreement was delayed by a further year; thus, 2000–2001 would mark the end of the implementation period.

"an unlikely hypothesis at the upper end of the scale" (Commission 1992, 8). In its view, the Mac Sharry reforms meant that "There is no longer any incentive for higher yields" (Commission 1992, 7). Others, however, have disputed these findings.

For example, Britain's Home-Grown Cereals Authority (1994, 10) has suggested that the Commission's estimates may be unduly optimistic. It suggests that

- the Commission's assumed growth in yields "could prove unrealistic given the continuous steady rise in cereal yields since the 1960s"
- there is a smaller potential for displacing cereal substitutes in animal feed rations than the Commission assumes
- "most projections of livestock numbers suggest a fall towards the end of the decade, hence reducing animal feed demand." (Home-Grown Cereals Authority 1994, 10 and 11)

There was bound to be some debate about the likely impact of the policy changes introduced into the cereals regime as a consequence of the Mac Sharry reforms, in part because of the political character of the issue and in part because the changes went well beyond the fluctuations in variables normally monitored in econometric estimations. The set-aside requirement will have some effect on the quantity supplied. But because of exemption from set-aside of small producers the overall area reduction will be much less than 15 percent, and in addition some policy slippage is bound to occur. The area compensation payments are being paid on such a generous scale that all farmers are expected to plant the maximum area to which they are entitled. Thus, the reduction in cereal prices will not result in an area response as would have been the case had the payments been truly decoupled. Economists do agree that the cut in cereal prices will result in a yield response as farmers apply smaller applications of fertilizer and agrochemicals to their crop and harvest it less carefully. But there is less certainty on the magnitude of the likely decline in yields. Josling's (1993, 99) guesstimate was that 70 percent of the overall supply response that economists would normally expect would be translated into this yield response. Thus, our expectation is that cereal yields will fall as a consequence of the Mac Sharry reforms, but not by a large amount. Productivity changes will continue to filter through into the cereal sector, and although productivity improvements are not necessarily reflected in an increase in yields—they might result in the use of fewer inputs to achieve the same output—it is our belief that for the next decade much of the change will be reflected in increased yields simply because the bulk of the new technologies have been developed in an environment of high land and cereal prices. Over time, lower input–output technologies will be developed for European conditions and adopted by European farmers. Thus, after a one-off reduction in yields brought about by the price cuts embodied in the Mac Sharry reforms, we would expect the upward trend in yields to resume. However, this will probably be lower than the yield increase of 2.5 percent per year during the 1980s (1.8 percent per year over the five-year period to 1992) reported by the Commission (1992, 7).

In 1993, the French government was particularly skeptical of the Commission's claims and, in the closing phases of the negotiations, sought to protect French farm interests from additional set-aside. For example, *Agra Europe* (17 December 1993, E2) reports that the European Council, meeting in Brussels in December 1993, had been asked by Germany and France to endorse the following declaration:

> The European Council has noted the declaration from the Commission, according to which the commitments taken on agriculture within the framework of the Uruguay Round are compatible with the reformed CAP. The European Council asks the Commission to monitor the impact of the GATT agreements on the agricultural policy and to report to the Council if the Commission's forecasts prove to be incorrect.
>
> In this case, the European Council invites the Commission to submit appropriate proposals taking account of the decisions taken by the Edinburgh European Council agreement on Community finances in order to ensure the implementation of the reformed CAP.
>
> In this context, the Council holds the view that:
>
> — In the arable sector: there will be no increase in the compulsory rate of set-aside nor any new quantitative limitation. Other measures could be taken instead.
> — In the beef sector: regardless of the review of the common market organisation, a guarantee to maintain the income of producers through a balanced adjustment of the level of premiums will have to be envisaged, including for extensive rearing.
>
> The European Council holds the view that it will be necessary to adapt the regional base areas in the new Laender to developments in current production while taking account of an increase of the order of 200,000 hectares of land resulting from the oilseeds agreement with the U.S. Besides, the Council holds the view that any

decrease in farm prices will have to be avoided during the marketing year which would result from a realignment and that such a decrease will have to be spread over several years.

The Council is invited to take the appropriate decisions.[13]

Rather than acceding to this request that Germany should be protected from realignments of the green mark and that the East German *Länder* should not be penalized for overshooting the regional base areas for arable area compensatory payments—as well as the French request that no further set-aside or other quantitative limits be imposed—the European Council instead adopted the following anodyne text:

> The European Council takes note of the Commission's prognosis of the compatibility with the reformed CAP of the new international commitments which would result from an agreement in GATT. If, however, additional measures were to prove necessary, the Council agrees that they should not increase the constraints of the reformed CAP, nor affect its proper operation. It would, if necessary, take the requisite steps while respecting the decisions of the Edinburgh European Council. (*Agra Europe,* 17 December 1993, E2)

On 2 September 1994, *Agra Europe* (P1), citing the just-published Home-Grown Cereals Authority's report we discussed earlier, deduced that "The main conclusion to be drawn from the Authority's analysis is that the choice lies between either a further drastic reduction in support prices . . . , or an increase in set-aside." A week later it was reporting that the French farm minister Jean Puech had met with Farm Commissioner René Steichen to ask for a reduction in the set-aside requirement for 1995–96 as a consequence of buoyant market circumstances for cereals and a reduction of EU intervention stocks (*Agra Europe,* 9 September 1994, P3). It would be interesting to know whether the French government was simply taking the opportunity to argue for a short-term expansion in EU farm production or whether it had revised its views on the longer-term impact of the Mac Sharry reforms.

---

13. It will be recalled from chapter 5 that the Edinburgh Meeting of the European Council in December 1992 (referred to in this quotation) had reaffirmed the decision to limit the growth in budgetary expenditure devoted to CAP support but had nonetheless signaled that monies could be found to meet additional expenditure stemming from the monetary disturbances experienced in the EMS.

CHAPTER 8

The Next Round: Toward Free Trade?

Throughout the post–World War II period, the normal GATT rules have had very little impact upon domestic farm policies and international trade in agricultural and food products. The world market has been in disarray. International prices have been depressed and destabilized, and the world's efficient agricultural producers have seen their markets captured by less efficient but highly subsidized growers from elsewhere. The EU's CAP has been a significant factor in generating this outcome, although other important traders such as the United States and Japan have also contributed to the disarray.

The Agreement on Agriculture concluded as part of the overall settlement of the Uruguay Round of GATT negotiations on 15 December 1993 and signed at Marrakesh on 15 April 1994 marks a significant change in the way countries can support their farm sectors and brings agricultural trade firmly within the GATT disciplines. The beneficial effect of the Agreement on Agriculture will gradually be realized over its six-year implementation period.

What is clear is that the agreement is fairly modest when contrasted with the aspirations allegedly held by some of the participants in the early stages of the negotiations. Protectionist trade barriers, export subsidies, and trade-distorting farm support will not be swept away. Instead, they are to be tightly constrained and modestly reduced as a result of the complex agreement negotiated. While it must be presumed that all signatories will respect the letter of the agreement, it must also be expected that most of them will wish to minimize the adverse impact on their rural communities and food and agricultural businesses, and much will depend on how the major trading powers actually implement the deal. It is still too early to predict what the overall impact will be.

Nonetheless, it must be expected that the prices of farm and food products on world markets will, on the whole, be higher than they would have been had the old policies continued unabated. The effect will be uneven: some world prices will rise by a greater percentage than others, and some may fall. On the whole, food-exporting countries will gain, but net cereal importers in the developing world may have difficulty adjusting.

Although the trade impact of the newly negotiated GATT Agreement on Agriculture could be dismissed as modest, this would misjudge the agreement's importance. In the agreement, the contracting parties recognize that "the long-term objective of substantial progressive reductions in support and protection resulting in fundamental reform is an ongoing process" and commit themselves to beginning a new round of multilateral trade negotiations "for continuing the process" before the end of the 1990s (Article 20). The debate on the priorities and objectives of that new round of negotiations will soon commence.

Within the EU, the CAP is under renewed attack. The "new" CAP relies more heavily on taxpayer support than did the "old" CAP prior to the Mac Sharry reforms of 1992, and this is likely to generate budgetary problems in the near future. Furthermore, the new system of support is far more transparent than its predecessor, and public reaction against massive CAP support payments to wealthy farmers and landowners is beginning to mount. CAP reform will again be on the agenda before the end of the decade.

With these considerations in mind, in this closing chapter we will attempt to

- outline the probable impact of the newly negotiated agreement on world trade;
- contribute to the debate on the objectives to be achieved in the next round;
- suggest further reforms for the CAP; and
- address the question, "Did the GATT negotiations influence CAP decision-making?"

Implications for World Trade

In attempting to model the potential impact of the Uruguay Round Agreement on Agriculture on future world market prices, it is tempting to take the "headline" figures of a 36 percent cut in import tariffs and a 21 percent cut in exports as the main explanatory variables. However, this would considerably overstate the case, for a number of reasons.

As we noted previously, the 36 percent cut in import tariffs is a simple arithmetical average, and for some key commodities a smaller reduction will be applied. In the EU, for example, we know that sugar import tariffs will be reduced by 20 percent. Furthermore, as table 8.1 indicates, significant import tariffs will remain even at the end of the six-year implementation period. Many of these tariffs will remain prohibitively high, and in conjunction with the application of the special safeguards provisions they will ensure that imports are not sufficiently competitive to increase their market share in some protected markets.

On exports, the 21 percent volume reduction refers to subsidized exports, and it will be possible—either as a consequence of an increase in world market prices or a shift in support to the EU's style of area compensation payments—for what were subsidized exports to continue in the future as unconstrained, unsubsidized sales. Enlargement of the EU will allow the EU to internalize some of its export sales, thus helping it to meet its reduction targets. For example, it had been suggested that the four nordic and alpine states that had planned to join the EU on 1 January 1995 would have an import deficit of 261 thousand metric tons of white sugar (*Agra Europe,* 29 July 1994, P4), and, if this had all been sourced from the EU, it would go a long way toward achieving the required reduction of 339.6 thousand metric tons per year in the EU's

TABLE 8.1.   Pre- and Post-Uruguay Round Tariffs
(Most Favored Nation [MFN] rates, after full implementation [percentage])

Importing Market	Unweighted MFN Tariff Average		Maximum MFN Rate	
	Pre-UR	Post-UR	Pre-UR	Post-UR
• Nontropical agricultural products				
Canada	13.4	8.6	266.6	170.6
EU	32.2	21.4	438.0	280.4
Japan	31.4	23.1	650.0	552.5
United States	10.3	7.2	999.8	449.9
• Tropical agricultural products				
Canada	4.9	2.5	30.1	19.3
EU	17.3	11.1	162.4	104.0
Japan	14.2	9.0	589.0	500.7
United States	9.5	6.5	536.9	429.5
• Textiles and clothing				
Canada	19.6	12.7	30.0	18.0
EU	10.5	8.2	17.0	12.0
Japan	10.7	6.9	22.4	16.0
United States	12.8	9.1	42.4	32.0
• Leather and footwear				
Canada	13.3	9.6	25.0	20.0
EU	8.3	6.9	20.0	17.0
Japan	29.8	23.4	84.0	75.2
United States	14.0	11.8	61.8	61.8
• All products (excluding fuels)				
Canada	9.9	5.5	266.6	170.6
EU	10.4	6.6	438.0	280.4
Japan	9.7	5.9	650.0	552.5
United States	7.0	4.3	999.8	449.9

Source: UNCTAD, *Trade and Development Report 1994,* as reported in the *Financial Times* (15 September 1994, p. 8).

subsidized sugar exports by the end of the transition period (Commission 1994, Annex V). Furthermore, the EU's commitment to reducing its exports of subsidized olive oil by 31.1 thousand metric tons per year by the end of the transition period could readily be met if Tunisia could be persuaded to sell its olive oil direct to the world market rather than take advantage of the preferential access arrangements it currently enjoys in the EU. Even if export constraints are effective, countries might nonetheless fail to curb production and instead allow intervention stocks to rise. These stocks, overhanging the world market, would undoubtedly depress world market prices.

With these caveats in mind, we show in table 8.2 the preliminary results of a modeling exercise undertaken by the Australian Bureau of Agricultural and Resource Economics (ABARE) using the Static World Policy Simulation (SWOPSIM) world agricultural trade model developed by the USDA. The figures show the percentage increase in world market prices that are likely to result from the Uruguay Round Agreement after full adjustment has taken place, that is, well into the next decade. We suspect, despite the efforts of the modeling team, that these numbers will overestimate the impact of the changes, for the reasons we outlined previously.

Some of these predicted changes are very modest, and a 1 percent increase over a decade is not notably different from no increase at all. None-

TABLE 8.2.   Likely Increases in World Market
Prices Resulting from Implementation of the
Uruguay Round Agreement

Commodity	Percent
Beef (FMD[a] free)	6
Beef (FMD affected)	1
Pork	7
Sheep meat	3
Poultry meat	2
Butter	4
Cheese	20
Milk powders	16
Wheat	8
Maize	6
Other coarse grains	5
Rice	8
Soybeans	1
Other oilseeds	6
Cotton	2
Sugar	1

Source: Andrews, Roberts, and Hester (1994, 70).
[a] FMD = foot-and-mouth disease

theless, most world market prices are expected to increase as a result of the agreement, although some—such as manioc and other cereal substitutes—could fall as a consequence of their protected export markets being undercut by cheaper imports. The agricultural sectors of countries like Australia will unambiguously gain as a result of higher export prices and enhanced export volumes. For the EU, the budgetary cost of export subsidies will be reduced, helping it to meet the GATT constraints more easily, and the overall beneficial impact on the EU economy will more than outweigh the minor losses suffered by the farm sector and food processors.

An important objective of the Cairns Group in the negotiations was to recouple domestic price movements to world market fluctuations. This should result in a dampening of world market price volatility. Tyers and Anderson (1992, 225) point out that what is required to achieve this effect is for countries to liberalize their agricultural trade policies or change existing policies so that only noninsulating instruments are used, such as ad valorem tariffs and export taxes and subsidies. Their model (which relates to 1980–82) suggests that if such a change were adopted by the industrial market economies alone, world agricultural price volatility would be reduced by one quarter. If all industrialized and developing countries adopted such policies, this "would reduce global food market volatility by more than two-thirds." Tariffication was an important first step in moving toward this goal. However, it is unlikely that the Uruguay Round Agreement will have the dramatic results predicted by Tyers and Anderson. This is because tariffication is to be applied only imperfectly, and many of the new tariffs will—in our judgment—remain prohibitively high. Furthermore, the export subsidy constraints do not limit the unit rate of subsidy paid, and thus export subsidies will remain variable. Prohibitively high tariffs, the special safeguards provisions, the import arrangements for cereals into the EU and rice into Japan, and variable export subsidies will all act to maintain the insulation of domestic markets from world market price movements.

With respect to the expected increase in world market prices, most concern focuses on those developing countries that are net importers of cereals, either on commercial or concessional (food aid) terms. As world cereal prices rise it is feared that developing countries will lack the foreign exchange to maintain their volume of commercial imports and that donors will be more reluctant to offer food aid.

It is not really possible to group all low-income countries in an analysis of this kind. The net effect will be country-specific. Some middle-income countries (such as Argentina and Thailand, which are, respectively, net cereal and rice exporters) will gain, as will net exporters of sugar, meat, and tropical products. For the net food importers, there will be an adverse macroeconomic impact on their economies as they adjust to higher import prices, and net food importers tend to be some of the least-developed countries in the world. Nonetheless, Pryke and Woodward (1994, 15) identify two moderating factors:

- Lower-income countries' food imports are typically more concentrated on basic staples (particularly cereals), whose prices are less affected by the GATT, while middle-income countries spend more on higher-value products (e.g., beef) whose prices may increase more.
- Grant food aid will tend to moderate the impact on the poorest countries, particularly in the cases of cereals and dried milk.

It is our view that the GATT constraints on exports will be binding for subsidized agricultural exporters such as the EU. Consequently, higher world market prices will not increase the opportunity cost to these regions of maintaining their food aid shipments. As now, food aid will prove to be a useful means for disposing of unwanted surpluses. It will be recalled that genuine food aid shipments are exempt from the GATT export constraints.

The GATT Agreement on Agriculture imposes no policy constraints on the less-developed countries, and for developing countries in general a ten-year implementation period is allowed to achieve two-thirds of the extent of the policy changes imposed upon the developed world. The governments of many developing countries intervene extensively in their food markets through direct taxes and subsidies and foreign exchange controls, thereby isolating their farmers and food consumers from world market prices. Often, the farm sector is heavily taxed. Thus, it is difficult to say with certainty what microeconomic effect will be felt by producers and consumers in the developing world as a consequence of the macroeconomic shocks their economies encounter in the face of higher agricultural prices on world markets. In some countries, however, higher world market prices might encourage governments to pursue greater liberalization of domestic food markets, which would encourage local farmers to expand production, thereby lessening the country's dependence on imports. The microeconomic benefits of such an outcome would tend to offset, and might even eliminate, the adverse macroeconomic impact of higher world food prices.

An Agenda for the Next Round of Negotiations

The three-part structure of the 1993 agreement—relating to import access, export subsidies, and internal support—has probably already determined the outlines of an agenda for the next round. While the Marrakesh declaration on Trade and the Environment promises to en-liven the agricultural negotiations, sanitary and phytosanitary matters may well come to the fore. The brief discussion that follows will focus upon the import taxes and domestic and export subsidies inherent in farm support.

During the course of the Uruguay Round Agreement, countries that fixed tariffs and bound the levels of domestic support and export subsidies in terms of currencies that subsequently experience rapid inflation will find the constraints to be more restricting than they had expected, and they may well lobby for some renegotiation of the constraints. This should be resisted, unless countries can show an excessively high rate of inflation experienced by the currency in which their constraints are written.

The present agreement reduces the average import tariff—as well as expenditure on export refunds—by 36 percent over a six-year period. If it is possible to envisage this momentum being maintained, two fur-ther agreements with six-year implementation periods, covering the years 2001–6 and 2007–12, could result in the total elimination of import tariffs and export subsidies in the developed world by the year 2012. The mathematics of this involves a 36 percent reduction over the period 1995–2000, a 50 percent reduction of the bound rate in the year 2000 over the period 2001–6, and a 100 percent reduction of the bound rate in the year 2006 over the period 2007–12. Table 8.3 and the following paragraphs outline a possible strategy.

Import Access

Although one might question the United Nations Conference for Trade and Development's (UNCTAD) methodology in determining the maxi-

TABLE 8.3. Summary of Preferable Elements of Future GATT Agreements

Implementation Period	Tariffs	AMS	Export Subsidies
1995–2000	−36 percent	−20 percent	−36 percent
2001–6	−50 percent −50 percent additional duties	−25 percent and redefinition of exempt payments	−50 percent
2007–12	elimination of border protection	−33.3 percent	elimination of export subsidies

mum ad valorem tariff rates recorded in table 8.1, it is clear from this table and our earlier discussions that for many agricultural products very high tariff rates will continue to apply even after full implementation of the Uruguay Round Agreement. Some of these tariffs will remain prohibitively high, and, consequently, imports will still be uncompetitive in these protected markets. The success of the Uruguay Round Agreement was that it forced countries to tariffy their border measures and reduced tariffs by 36 percent on average. But now that those tariffs are determined, trading partners can readily ask for further reductions in subsequent negotiations. The limitations of the agreement were that (1) tariffication resulted in very high tariffs, some with considerable water in the tariff, such that imports remained uncompetitive, and that (2) provided each tariff was reduced by a minimum of 15 percent, countries could choose how they might achieve the average reduction of 36 percent.

Based upon the precedent of 1993, it might be hoped that over a new six-year implementation period a similar reduction of the year 2000 tariff levels could be negotiated. As we explained in the previous section, mathematically this requires a 50 percent reduction in the tariff rates of the year 2000 in order to achieve the same effect as a 33.3 percent reduction in the tariff level of the base period. However, in this case it would be preferable to seek a larger percentage reduction in the highest tariff levels. But failing this, an across-the-board reduction of 50 percent should be sought. A third agreement, covering the six-year period 2007 to 2012 could then see the complete elimination of tariffs. Given the earlier experience of tariff reductions for manufactured goods, this should be achievable.

In 1993, in an attempt to achieve minimum access arrangements, some tariff quotas were agreed upon. The politics of this decision are easy to understand—some import access had to be secured—but the decision was flawed. Although it is too early to judge how the arrangements will work in practice, there is a very real danger that the main beneficiaries will be the traders lucky enough to secure the associated import licenses for the protected markets or the civil servants who issue the licenses. They will capture additional profits in the form of quota rents. The sooner these unearned quota rents are driven from the system, the better. They will disappear either when tariff quotas are no longer prohibitive, so that traders can ship all they wish to import at the within tariff-quota tariff rate, or when the MFN tariff rate is reduced to the level of the tariff-quota tariff rate. Thus, where it can be shown that existing tariff quotas are fully taken up, the next agreement should include a clause that insists upon a marked and consistent increase in the quantity specified in the tariff quota or a more substantial cut in the MFN tariff rate.

The special safeguards provisions in the existing agreement pose a more technical issue that will have to be dealt with in the next negotiations. In particular, this involves the reference price that allows countries to impose an additional duty on a consignment basis if the current CIF import price falls below its 1986–88 average level. There are two questions to be dealt with here: first, should there be a revision of the reference period, and, second, should not more control be exercised over the determination of the reference price? We saw in chapter 7 that the EU, on the basis of the existing agreement, has set reference prices that will result in frequent application of the safeguards clause. And we suspect that the EU's trading partners will wish to review this in the next round. It might also be argued that a more recent period would be more relevant for determining the reference price. But, if world market prices had risen after 1986–88, this use of a more recent period for the reference price would mean that the special safeguards provisions could be triggered more easily. There should also be a 50 percent reduction in the level of additional import duties that can be applied. This step, together with the further reduction in tariffs we proposed earlier, would allow the recoupling of world and domestic markets and would significantly reduce the instability of world prices.

Domestic Support

For the EU, and we suspect for a number of other countries as well, the 1993 agreement to reduce the overall level of farm support by 20 percent as measured by an AMS, coupled with the decision to exempt a wide category of "direct payments under production-limiting programs" from the commitment, meant that this element of the agreement is likely to have little practical effect. However, this outcome would not be sufficient justification for abandoning the concept of reducing domestic support in the next agreement.

A further reduction in the AMS should be agreed to, but mathematically it needs to be a reduction of 25 percent of the new base in order to achieve the same momentum of change. More controversially, we would suggest that the Cairns Group should seek to reopen the debate over exempt payments. As we pointed out earlier, it cannot be claimed that the EU's area compensation and livestock headage payments are production-neutral and thus have no impact upon world trade or that they satisfy the legitimate interests of the EU. They should be classified into the amber box and be subject to reduction in the next agreement. It should be noted that, provided these area compensation and livestock headage payments are not classified as export subsidies, this proposal would allow these payments to continue well into the twenty-first century. Later in this chapter, we will argue that it would be in the EU's own interest to eliminate these payments well before then.

Export Subsidies

In the EU it is the export constraints, and in particular the 21 percent reduction in the volume of subsidized exports introduced by the GATT agreement of 1993, that cause the most concern. The agreement provides for a 36 percent reduction in expenditure on export subsidies. Following the earlier logic, of the last section, we would suggest that in the next agreement the contracting parties should agree to a 50 percent reduction in expenditure on export subsidies over six years from the year 2000 base achieved in the present agreement. To achieve mathematical consistency, the reduction in the volume of subsidized exports should be 27 percent of the year 2000 base.

Trade and the Environment

At Marrakesh in April 1994, ministers acknowledged that in the coming years environmental concerns had to be recognized and resolved in the WTO. There are two levels of concern. First, many environmentalists believe that a more liberalized world trading system will lead to faster rates of economic growth and that economic growth inevitably leads to environmental degradation. It will require considerable pedagogical skill to assuage this fear and considerable diplomatic skill to convince many people in low-income countries that the West's advocacy of environmental controls is not a capitalist ploy designed to deny them the living standards enjoyed by their Western counterparts.

But there are also more technical concerns that must be addressed. Many countries now subscribe to the "polluter pays" principle, in which the costs of pollution are internalized into the operating expenses of the polluting industry, or, alternatively, the industry is taxed for the pollution caused or subsidized for the beneficial externalities generated. If imported products enjoy a cost advantage because they do not face the same pollution controls or taxes, then home producers are bound to feel aggrieved in the face of what they perceive to be "unfair" competition. Lax pollution controls in competing regions may however simply reflect the comparative advantage those regions enjoy, either because of lower densities of productive activity, which allow a given pollutant more easily to be dissipated in the environment, or because of a higher tolerance in the local population to given levels of pollution. Similarly, producer interests in exporting countries might well object to subsidies in the importing region that are ostensibly paid to achieve environmental objectives but that in fact help subsidize local production. If farmers help to set the agenda for determining environmental subsidies, can we be certain that payments for the maintenance of dry stonewalls do not in fact subsidize sheep production?[1]

Increasingly, environmental characteristics are incorporated into the product itself. Thus, concerns about animal welfare do not simply extend to animal husbandry techniques in the country concerned but also to the methods of production of all livestock products sold in the country. In addition, legislation relating to packaging and recycling applies equally to imported and domestically produced goods. Thus, environmental legislation inevitably impinges upon trade and gives rise to two conflicting fears:

- that GATT rules will forbid the effective application of environmental laws that seek to protect legitimate domestic environmental concerns; or,
- that under the camouflage of environmental protection, "green" nontariff barriers will be put in place that thwart the legitimate trading interests of low-cost suppliers.

Clearly, the battle lines are drawn, and these opposing views will lead to much debate in the coming years.

Priorities for Further Reform of the CAP

The 1992 Mac Sharry reforms can be criticized on two counts: (1) they will be extremely burdensome to the EU's budget, and (2) they are not entirely production-neutral and, in particular, will sustain rural land prices within the EU. The first of these criticisms stems directly from the switch from consumer to taxpayer support, and the fact that—as we explained in chapter 5—the Commission's original proposal that compensation payments be modulated was rejected by the Council of Ministers. Modulation—a piece of Eurojargon—implied full compensation of

---

1. A thought prompted by a seminar presentation by Deborah Gourlay of the University of Aberdeen.

revenue loss to small disadvantaged producers but only partial compensation for larger farmers. Budgetary crises in the latter part of the 1990s are likely to result in a review of these compensation arrangements. Indeed, the outgoing farm commissioner, René Steichen, intimated that, had he continued to serve in the 1995–99 Commission, he would have wished to see this "flaw" in the arrangements reversed by way of a determination of a maximum payment per producer (*Agra Europe,* 9 September 1994, P4). Alternatively, the Council could decide that the payments would be phased out over a specified period of, say, ten years. Under present legislation, the payments will be made forever or until the Council changes the provisions.

Area compensation payments are tied to the continued cropping of farmland. It is to be expected that yields will fall, reflecting the lower price that cereals fetch in the marketplace as lower applications of fertilizers and agrochemicals are applied and harvesting effort is reduced. However, as we explained in chapter 7, it is unlikely that the total area devoted to arable crops will fall as a result of the changed system of support—although the set-aside requirement will of course have an effect. Hence, the cereal price reduction inherent in the Mac Sharry reforms will have only a muted impact on output: there will be a yield response but not an area response. Thus, it cannot be claimed that the EU's area compensation payments, or indeed its livestock headage payments, are entirely production-neutral, despite the fact that in the 1993 GATT agreement they were deemed to be exempt from the domestic support reduction commitments.

It can hardly be claimed that area payments are made in *compensation* for loss of revenue because they are tied to the continued cropping of farmland and are payable to anyone who crops eligible areas of farmland, whether or not the claimants are recent or early entrants to farming. Moreover, it is certain that area payments will sustain rural land prices. Farmland prices are notoriously difficult to model. The price of a particular piece of land will be influenced by interest rates and expectations about inflation and the movement of land prices as a hedge against inflation. It will also be influenced by its residential and sporting value, any economies of scale that amalgamation with an adjoining holding may bring, and the expectation of enhanced land values if the land were to be rezoned for urban development. Nonetheless, farming profits undoubtedly play a role: the higher the return from farm production, the greater the land price is likely to be, other things being equal.

Thus, the rent that landlords will demand of arable land as well as its market price will undoubtedly reflect the market's expectation of a continued flow of area compensation payments. Inflated land prices will mean that alternative crops, and other rural land uses such as forestry or amenity, will be placed at a commercial disadvantage unless they too are subsidized at the same rate as arable crops. Land-saving technologies such as increased fertilizer and agrochemical applications will be encouraged on these alternative land uses, and financial incentives will remain to convert "nonproductive" woodland, wetlands, scrub, and hedgerows into "productive" agricultural use. Thus, the environmental degradation engendered by the CAP will continue.

An alternative policy mechanism is at hand that will not result in these unfortunate consequences. European agricultural economists have long advocated the replacement of existing support mechanisms by truly decoupled compensation payments. Tangermann (1991), for example, is a recent advocate of such a scheme. Our suggestion is that the existing area compensation and headage payments, and similar compensation schemes that will apply in the future to other sectors as yet untouched by CAP reform, be converted into a simple bond scheme.

All recipients of aid on vestment day, say, 31 December 1994, would receive a bond that—in its first year of validity—would entitle the holder to receive a payment equal to the total aid that the farmer would have received under the previous regime. Payments would continue over a ten- or fifteen-year period, reducing annually, and then cease. Payments would be made to the owner of the bond and would not be conditional upon a continuation of farming. Thus, farmers could retire and still receive the flow of compensation payments. They could sell the bond and thus receive a capital sum representing the market's present valuation of that future income flow, and on their deaths their heirs could inherit the bond.

There are two major advantages of this scheme. First, the capital value associated with CAP support would be stripped out of land and other fixed asset prices and transferred instead to the bond. Thus, new entrants to the industry would not have to pay an absurdly high entry price to buy into CAP largess. Instead, their overhead costs would reflect world prices for farm products. As land prices fell, then other land-use activities, such as forestry, could more readily take up land that was no longer commercially viable in farming. Second, there would be greater certainty about the future flow of support payments. Under the present system, the farm sector is rightfully concerned that CAP support payments could be substantially reduced, or eliminated, at any moment on the whim of the Commission and Council of Ministers. Bonds, however, once issued and with payment terms clearly stated and guaranteed by the finance ministries of the member states, would rank with other gilt-edged stock as secure investments.

The aim of this scheme is to ensure that those who suffer capital loss as a consequence of the removal of farm price support receive appropriate financial compensation to forestall the wave of bankruptcies that would otherwise sweep through the farm sector and rural community. But therein lies the major drawback to the scheme. The bonds would be issued to "farmers," however defined. If these individuals own the land, mortgaged or unencumbered, no problems emerge. If, however, tenant farmers are compensated for the capital loss suffered by landlords, then the scheme would be misplaced, and consequently some mechanism for sharing the compensation between tenant and landowner would have to be agreed on—perhaps along the lines of schemes adopted in the United Kingdom to allocate the value of milk quotas between tenant and landowner on sale of quota. Farmworkers—under any adjustment scheme—would receive no compensation for their lost earning opportunities, nor would the associated input suppliers and first-stage food processors.

A bond scheme would genuinely be a decoupled income support measure and would thus be exempt from any reduction commitments under the Uruguay Round or any future GATT agreement. Despite these obvious advantages, a bond scheme remains deeply unpopular in political circles. Nonetheless, the proposal still has some support. In his recent book on the future of Europe, Sir Leon Brittan (1994, 130)—who is serving as a commissioner in the 1995–99 Commission—wrote approvingly of the benefits of such a scheme, declaring that it "would help heal the sclerosis currently gripping Europe's agricultural markets." Perhaps the new Commission will take up the challenge.

Institutional Arrangements

In the latter half of the 1990s, the EU will undergo a number of institutional changes that will affect its capacity and willingness to undertake CAP reform. A new college of commissioners took office on 1 January 1995 and will serve until 31 December 1999. It is these individuals, and in particular the as yet unknown commissioner for agriculture, Franz Fischler, who will be responsible for proposing policy reform to cope with the budgetary crises of the late 1990s, the launch of a new round of GATT negotiations focused on further reductions in farm support, and continued negotiation with central and eastern European states seeking improved access for their food and drink exports to the EU and eventual EU membership.

Austria, Finland, and Sweden joined the EU on 1 January 1995. Exactly how fifteen farm ministers will interact in the Agriculture Council is far from certain. Sweden in recent years has undertaken significant agricultural reform and might be expected to join the reformist camp. On the whole, however, these new member states have had highly protected farm sectors, and their farmers have protested against the lower levels of support they will receive under the CAP. Our fear is that the new fifteen-member Farm Council will be instinctively more supportive of the farm sector than its predecessor.

Decision making under the qualified majority rules in a fifteen-member Council will be even more difficult than it is today. In 1996, a new intergovernmental conference is to be convened that must address this issue, among others. Already the debate has begun. Will this debate result in an inner core of committed federalists pressing ahead with integrationist policies including the CAP, with an outer ring of less closely associated states in a free trade area? Or will it result in a Europe of "variable-geometry" in which different coalitions of member states pick and choose "common" policies and in which some might choose not to apply the CAP? Or, alternatively, will we retain a single grouping of all member states applying most policies in common and pursuing the European experiment at the speed dictated by the most reluctant member? And, if the latter scenario is the most likely, what—if any—changes will be made to the rules on qualified majority voting, and what will be their practical outcome in terms of CAP reform?

All are agreed that a further expansion of the EU to include certain central and eastern European states as well as Cyprus and Malta must await the outcome of the 1996 intergovernmental conference. And all are agreed that the existing CAP could not survive this enlargement. The agricultural potential of the prospective central and eastern European member states is such that—with the present arrangements—surpluses and budget expenditure would soar. However, there is as yet no consensus on a strategy for CAP reform. The extension of the present area compensation payments to farmers in new member states who suffered no income loss in the Mac Sharry reforms of 1992 is a nonsensical notion. It would be far better to introduce tradable bonds now and thus move CAP support prices closer to long-term world market prices. Failure to do so would encourage governments in the potential new member states to adopt support policies that mimic the present CAP in the unjustified belief that, when they become members of the EU, their farmers will be funded from an enlarged EU budget.

GATT and CAP Reform

Institutional structures do condition the framework in which policy decisions are made. It is our firm belief that the present institutional arrangements have been inimical to radical CAP reform for many years. It is too easy to block a reform proposal and too difficult to stop the CAP from rolling forward unchecked over the decades. Prior to 1992, significant policy change had only ever been triggered by budgetary crisis.

Individuals do matter. Mansholt, Lardinois, and Mac Sharry are remembered because their personalities helped shape the CAP. Political pressures are also important. The French government, for example, does consider the electoral significance of the farm vote, and it does believe that the farmers' street protests influence the opinions of the electorate in general.

The farm lobby, backed by associated interests in the input supply and first-stage processing industries, does remain more influential than the consumer lobby, and understandably so. When a small coherent group has a very tangible vested interest to defend, it will be far more willing to fund an effective lobbying organization than would a much larger, less coherent group with only an indirect and individually small financial interest in the policy outcome. In recent years, the environmentalist lobby, broadly defined, has grown in importance and seems set to have more influence in setting the future agenda on farm policy.

In short the policy-making arena is complex and difficult to understand, and the outcome of the policy-making process is uncertain. Against this background, what can be said about the importance of external factors, and in particular the Uruguay Round of GATT negotiations and the progress of CAP reform?

We can find very little evidence to suggest that, when in September 1986 in Punta del Este, the EC embarked upon the negotiations, it had any real expectation that significant change to the CAP would result. Some individuals—who were influential at the time—did, but they were in a minority, and their grasp on the levers of power would not last long enough to see the negotiations through to a successful conclusion. The negotiations lasted longer than they should have in part because of a gross miscalculation on the part of the U.S. administration in its facile pursuit of the zero option and in part because of the unwillingness of the Commission and the majority of member states to outvote France in 1993 after most member states had accepted the inevitability of adopting the Blair House Accord.

The Mac Sharry reform proposals of 1991 were prompted by the struggles in the Council to adopt a credible negotiating mandate in October 1990 and, in particular, by Mac Sharry's promise to Germany that farmers would be compensated for the cuts in support prices implicit in the Commission's proposal. Thus, we conclude that the Uruguay Round GATT negotiations did impact on CAP reform, despite the Commission's insistence at the time that these reforms were only internally driven.

Some of the radical elements in the Mac Sharry reform proposals were emasculated by the opposition of a minority of member states, led, ironically, by the British farm minister John Gummer. The package accepted in May 1992, however, did mark a real break with the past. The package of measures was not totally compatible with Arthur Dunkel's Draft Final Act of December 1991. In particular, the area compensation payments could not be described objectively as production-neutral. Thus, the CAP reform package of May 1992 influenced the shape of the final GATT agreement in that at Blair House in November 1992 the two main parties to the agricultural negotiations—the United States and the EC—agreed between themselves that a wider range of direct payments to farmers, including the Mac Sharry payments and deficiency payments in the United States, would be exempt from the domestic support constraints.

After Blair House, the Commission insisted that the emerging GATT deal was consistent with the reformed CAP, once all sectors had been reformed. To some extent, this claim was justified. However, the EU did commit itself to tariffication and a range of constraints on maximum tariff rates and export subsidies that will have some impact on the shape of the CAP in future years. It is our view that the EU felt it had no choice but to accept this outcome. The tariffication and limits on maximum tariff rates are not a set of constraints that EU farm ministers would have willingly voted to impose upon the CAP. Rather, they are a set of constraints that, in the final analysis, the member states were able to assume because of the other benefits they expected their economies to reap from the overall agreement. Thus, we do concede that seven years of tortuous diplomacy had brought some change to the CAP, but this fell far short of the initial aspirations of the Cairns Group.

The main legacy of the Uruguay Round negotiations in terms of CAP reform is that a system of direct payments to producers was introduced in May 1992. The transparency of these payments, the size of the transfers to individual farmers, and the budgetary cost of the policy could, we believe, result in a coalition of reform-minded governments that would be willing and able to push through a bond scheme of the sort we advocated in the previous section. This is particularly true if the 1996 intergovernmental conference decides to simplify the qualified majority voting procedures.

Furthermore, agricultural trade is now subject to GATT disciplines. The exemptions and derogations from the normal GATT rules that we outlined in chapter 1 no longer apply. In particular:

- the process of tariffication, although imperfectly applied, was approved. Earlier GATT rounds were more successful in reducing trade barriers on manufactured goods because tariffs are both more transparent in their effect and intrinsically easier to handle in tariff negotiations than are variable import levies and other nontariff trade barriers; and,
- although of limited impact in the first instance, the import access and export subsidy constraints introduced in the Uruguay Round Agreement do define the starting point for the next round of negotiations. As such, they do offer the prospect of a removal of all trade-distorting support to the farm sector in developed countries over a twenty-year period.

But we should caution that despite many prophesies of its imminent demise the CAP has survived all earlier attempts at radical reform. We will watch with keen interest over the coming years.

Postscript

This text was completed in September 1994, but publication was delayed by several months due to difficulties beyond the control of the authors or of the Trade Policy Research Centre. This postscript records the major developments between October 1994 and April 1995.

EU Enlargement

The EU was enlarged on 1 January 1995 to embrace three of the four candidates that had negotiated entry: Austria, Finland, and Sweden. First in 1972 and then again in 1994 the Norwegian electorate voted against EU membership in a referendum, on the more recent occasion by a 52.2 percent to 47.8 percent margin (*Financial Times*, 30 November 1994, p. 16). Despite clear majorities in the other three states, eleventh-hour brinkmanship continued to dog the enlargement. In particular, in an attempt to resolve a long-standing dispute of Spanish access to fisheries around Ireland, Spain had threatened to veto the accession. The threat was only withdrawn on 22 December 1994 (*Financial Times*, 23 December 1994, p. 18).

Enlargement raises the question of compensation, under GATT Article XXIV:6, for other GATT signatories that lose as a consequence of the three newcomers' adopting the EU's common external tariff. It is the EU's view that, on balance, trading partners will gain as a result of the generally higher tariffs of the three new member states adjusting downward, and thus no compensation will be required (*Agra Europe*, 10 March 1995, P3). The EU's trading partners do not necessarily accept this view, arguing that compensation should be granted to countries that face tariff increases on individual items. A second GATT matter yet to be resolved relates to the aggregation of the EU–12's commitments under the Agreement on Agriculture with those of the three. If, for example, export commitments of the twelve and the three are simply added together, then what were subsidized exports from the twelve to the three now become internal EU trade, enabling the EU–15 to achieve its reduction commitments more easily.

In the new Commission, led by Jacques Santer, the Austrian Franz Fischler has been appointed commissioner for agriculture, and Sir Leon Brittan, while losing some of his responsibilities, has retained the trade portfolio dealing with GATT and the WTO. In the Council of Ministers of the EU–15, a blocking minority is now twenty-six votes:

Belgium	5
Denmark	3
Germany (Deutschland)	10
Greece (Ellas)	5
Spain (Espana)	8
France	10
Ireland	3
Italy	10
Luxembourg	2
Netherlands	5
Austria (Osterreich)	4
Portugal	5
Finland (Suomi)	3
Sweden (Sverige)	4
United Kingdom	10
Total	87
Qualified Majority	62
Blocking Minority	26

The WTO

The new World Trade Organization came into being on 1 January 1995, initially with thirty-eight members including the EU and its fifteen member states. All 128 members of the 1947 GATT are automatically entitled to become members of the WTO on formal acceptance of the Uruguay Round Agreements, and about twenty other countries are negotiating membership (*WTO Focus,* no. 1, January–February 1995). Peter Sutherland temporarily assumed the role of director-general, as a unanimous agreement could not easily be reached on which of the three candidates should serve. After the United States' candidate, the Mexican Carlos Salinas, had withdrawn and the Korean Kim Chul-su had stood down, the United States reluctantly agreed to support the EU's candidate. Thus, the Italian Renato Ruggiero embarked upon a four-year term as director-general on 1 May 1995.

All the main players in the GATT negotiations had completed their formal ratification procedures by 31 December 1994. In both the EU and the United States, however, this had involved elements of drama. In the EU, the European Court ruled in November 1994 that the European Union had executive competence to enter into agreements dealing with trade in goods but that the EU and the member states shared responsibility on other matters covered by the Uruguay Round Agreements (*Financial Times,* 22 November 1994, p. 20). This, as well as the overwhelming endorsement of the European Parliament (*Financial Times,* 15 December 1994, p. 5), removed the final obstacles to the EU's ratification of the agreements. In the United States, procedural delays meant that Congress had to be recalled after the mid-term elections held in November. But eventually both the House of Representatives and the Senate voted in favor (*Financial Times,* 2 December 1994, p. 1).

The EU's Implementation of the Agreement on Agriculture

In April 1995, there still remained considerable uncertainties for traders relating to the detailed application of the Agreement on Agriculture. The GATT agreements came into force on 1 January 1995, but for most agricultural products the new trade arrangements did not apply until 1 July 1995. Nonetheless, imports of tomatoes, courgettes, and cucumbers into the EU became subject to the controversial new minimum import price regime on 1 January 1995 (see Swinbank and Ritson 1995).

As far as exports are concerned, the Council has decided that prefixed export licenses will be required for all subsidized exports. Prefixing of export licenses is a long-standing aspect of EU policy: in essence, on lodging a financial security, a trader is granted an export license valid for a specified period and on which the export refund is already determined.[1] A failure to make use of the license would result in the forfeiture of the deposit. This system allows the EU to exercise some control over exports. By April 1995 the trade knew that existing prefixed export licenses would lapse on 30 June 1995, whatever their period of validity, but they did not yet know the details of the new system that would be in force on 1 July. Nor was it clear how the EU would ration the allocation of export licenses if and when the GATT limits became binding. *Agra Europe* reported that the delay was due in part to differences within the Commission services: DGVI (Agriculture) wanted to count the volume of licenses issued after 1 July 1995 against the EU's GATT commitment,

---

1. Valid, say, for the month of issue as well as the following three months.

while DGI (External Relations) wanted to monitor the exports actually shipped after 1 July (*Agra Europe,* 31 March 1995, p. 4).

For imports, the EU's additional commitment on limits to the import charge levied on cereals continued to cause concern. It will be recalled that, for cereals, the EU had agreed that it would undertake "to apply a duty at a level and in a manner so that the duty-paid import price for such cereals will not be greater than the effective intervention price . . . increased by 55 percent" (Commission 1994, headnote 6 to section I-A). If applied on a consignment basis, traders would have an incentive to declare a consignment value that would result in a zero import tax. If applied on a uniform basis it would result in a reinvention of variable import levies, and the EU could not easily guarantee that the commitment could be respected for each consignment.

In December 1994, the Council decided that the appropriate import charge would be determined for six categories of cereals rather than by fixing the import charge on each consignment. In reality, this means that a system akin to the old variable import levies is to be retained for cereals. The intent is to set six separate import tariffs every two weeks, for three qualities of common wheat (high, medium, and low), durum wheat, maize, and barley. For each, an indicative variety and market has been proposed. Canada, however, has pointed out that in 1993–94 it supplied 68 percent of the EU's wheat imports and that the main require-ment of U.K. millers—Canadian Western Red Spring wheat—usually commands a price premium over the indicative variety the EU plans to monitor when determining the appropriate import tariff for high-quality wheat. A similar problem relates to malting barley. Thus, before it has begun, the new system stands accused of consistently imposing import tariffs on certain varieties of cereals that result in the EU's GATT com-mitment being breached on a consignment basis (*Agra Europe,* 7 April 1995, P3). Unless a political accommodation can be reached, Canada and the EU appear to be on a collision course for a new trade dispute in the WTO.

Set-Aside

Although studies of the compatibility of the Mac Sharry reforms with the GATT export constraints continue to suggest that EU cereal produc-tion by the late 1990s will be excessive, under pressure from the French farm lobby, the Commission and Council have adopted amendments to set-aside that reduce the severity of the policy. In particular, faced with higher than expected EU market prices for cereals, the EU reduced the rotational set-aside requirement for the 1995 cereal crop from 15 to 12 percent of the arable area (*Financial Times,* 26 October 1994, p. 35). It was feared that high market prices would reduce the incentive to increase the use of cereals in animal feed. But if this had been the EU's real concern, lower cereal prices could more readily have been engineered by a reduction of export subsidies and import levies.

Green Money

On 1 February 1995 the green ecu used in the EU's CAP and the switchover mechanism were both abolished. This meant that all green conversion rates, used to convert CAP prices expressed in ecu into national currency, were divided by the coefficient 1.207509, and all CAP prices expressed in ecu were multiplied by the same coefficient. In national currency terms, of course, this left CAP prices unaltered. However, CAP prices continue to be fixed in ecu, and a green conversion rate is still required to convert from ecu into national currencies.

As the international values of EU currencies vary, it is necessary that green conversion rates be revalued and devalued to maintain comparability between green conversion rates and actual exchange rates. In abolishing the switchover mechanism, however, the Council of Ministers maintained an asymmetry in the system. Green conversion rate devaluations are triggered much more easily than revaluations, and with renewed currency instability in the EMS this has again become a matter of major debate in European farm policy circles. By April 1995, five countries (Belgium, Luxembourg, the Netherlands, Germany, and Austria) were resisting revaluations of their green currencies, which would trigger reductions of one or two percentage points in support prices in national currency terms.

A critical date in the annual calendar is 1 July, for it is the green conversion rate valid on that date that is used to determine the national currency value of area payments. Thus, farmers in countries that face green conversion rate revaluations would prefer to see such revaluations deferred until after 1 July. The willingness of the Council of Farm Ministers to pander to these concerns does not augur well for further reform of the CAP.

The new green money system, agreed upon in December 1994, does however provide a further twist. If a green conversion rate is revalued and such a revaluation does not simply reverse earlier devaluations, then ecu prices can be increased to offset any reduction in national currency terms generated by the revaluation. This provision has not yet been invoked, but it would have a clear inflationary impact as farmers in all countries would benefit from the increase in ecu prices, not just those individuals located in the revaluing country. If this principle were extended to area compensation payments, then a potential conflict with the GATT agreement could arise. It will be recalled that the peace clause only exempts direct support payments "provided that such measures do not grant support to a specific commodity in excess of that decided during the 1992 marketing year" (GATT 1994a, Article 13.2[b]).

Concluding Comment

It is still far too early for a clear picture of the impact of the Uruguay Round on world trade to have emerged. As we indicated in chapter 8, despite the headline figures of a 36 percent reduction in tariffs, a 20 percent reduction in domestic support, and a 21 percent reduction in the volume of subsidized exports, the Agreement on Agriculture is likely to have only a modest impact on world prices by the turn of the century because of the way in which the agreement is to be implemented. For example, the method of calculating tariff equivalents by the EU; the choice of the 1986–88 base period; and the exemption of EU area and headage payments, U.S. deficiency payments, and the costs of maintaining intervention stocks from calculations of domestic support will considerably reduce the final impact of the agreement.

Nevertheless, the Uruguay Round is likely to be seen as having made a significant contribution to the process of agricultural trade reform in the longer term. The successful conclusion of the multilateral trade negotiations resulted in the establishment of principles that are likely to lead to further liberalization of agricultural trade in future rounds. Agricultural trade has, for the first time, been brought within the rules and principles of GATT trade; the conversion of nontariff barriers to tariffs will increase the transparency of agricultural protection; and there has been recognition of the need to adopt support mechanisms that have less distortionary effects on agricultural markets. The acceptance of these principles together with the modest reductions in protection levels already achieved provide a firm basis for future liberalization.

April 1995

References

Agra Europe. Various issues. Tunbridge Wells, UK: Agra Europe (London) Ltd.

Agra Europe. 1993. *The GATT Uruguay Round Agreement—An Agra Europe Special Supplement.* Tunbridge Wells, UK: Agra Europe (London) Ltd.

Andrews, N., I. Roberts, and S. Hester. 1994. "The Uruguay Round Outcome: Implications for Agricultural and Resource Commodities." In *Outlook 94* Vol. 1, *World Commodity Markets and Trade.* Proceedings of the 1994 Outlook Conference held by the Australian Bureau of Agricultural and Resource Economics, 1–3 February 1994. Canberra: Australian Bureau of Agricultural and Resource Economics.

Ansell, D. J., and S. A. Vincent. 1994. *An Evaluation of Set-Aside Management in the European Union with Special Reference to Denmark, France, Germany, and the UK.* Centre for Agricultural Strategy Paper No. 30, Centre for Agricultural Strategy, The University of Reading, Reading, UK.

Australian Bureau of Agricultural and Resource Economics (ABARE). 1988. *Japanese Agricultural Policies.* Policy Monograph No. 3. Canberra: Australian Bureau of Agricultural and Resource Economics.

Baldwin, R. E. 1987. "Multilateral Liberalization." In *The Uruguay Round: A Handbook on the Multilateral Trade Negotiations.* Edited by J. M. Finger and A. Olechowski. Washington, D.C.: The World Bank.

Brittan, L. 1994. *Europe: The Europe We Need.* London: Hamish Hamilton.

Bureau of Agricultural Economics (BAE). 1981. *Japanese Agricultural Policies: Their Origins, Nature and Effects on Production and Trade.* Policy Monograph No. 1. Canberra: Australian Government Publishing Service.

———. 1985. *Agricultural Policies in the European Community: Their Origins, Nature, and Effects on Production and Trade.* Policy Monograph No. 2. Canberra: Australian Government Publishing Service.

Carson, Rachel. 1963. *Silent Spring.* London: Hamish Hamilton.

Changes to the Draft Final Act Required by US/EC Blair House Agreement. Unpublished, undated, and unattributed document made available by the Commission of the European Communities.

Commission of the European Communities. Monthly. *Bulletin of the European Communities.* Luxembourg: Office for Official Publications of the European Communities.

———. Monthly. *European Economy, Supplement A.* Luxembourg: Office for Official Publications of the European Communities.

———. 1968. *Memorandum on the Reform of Agriculture in the European Economic Community*. COM(68)1000, Part A. Brussels: Commission of the European Communities.

———. 1975. *Stocktaking of the Common Agricultural Policy*. COM(75)100. Brussels: Commission of the European Communities.

———. 1980. *Reflections on the Common Agricultural Policy* COM(80)800. Brussels: Commission of the European Communities.

———. 1983. *Common Agricultural Policy: Proposals of the Commission*. COM(83)500. Brussels: Commission of the European Communities.

———. 1984. *The Agricultural Situation in the Community: 1983 Report*. Luxembourg: Office for Official Publications of the European Communities.

———. 1986. *The Agricultural Situation in the Community: 1985 Report*. Luxembourg: Office for Official Publications of the European Communities.

———. 1987a. *The Agricultural Situation in the Community: 1986 Report*. Luxembourg: Office for Official Publications of the European Communities.

———. 1987b. *Report from the Commission to the Council and Parliament on the Financing of the Community Budget*. COM(87)101 and corrigendum. Brussels: Commission of the European Communities.

———. 1989. *The Agricultural Situation in the Community: 1988 Report*. Luxembourg: Office for Official Publications of the European Communities.

———. 1991a. *The Development and Future of the CAP*. COM(91)100. Brussels: Commission of the European Communities.

———. 1991b. *The Development and Future of the CAP. Follow-up to the Reflections Paper, COM(91)100 of 1 February 1991*. COM(91)258/3. Brussels: Commission of the European Communities.

———. 1992. *Agriculture in the GATT Negotiations and the Reform of the CAP*. SEC(92)2267. Brussels: Commission of the European Communities.

———. 1993a. *The Agricultural Situation in the Community: 1992 Report*. Luxembourg: Office for Official Publications of the European Communities.

———. 1993b. *Recommendation for a Council Decision Concerning the Conclusion of an Agreement on Certain Oilseeds between the European Economic Community and the United States of America within the Framework of the GATT*. SEC(93)53. Brussels: Commission of the European Communities.

———. 1994. Annex 1 of the European Community Schedule. Pt. 1. Most-Favoured-Nation Tariff, Section I Agricultural Products. Brussels: Commission of the European Communities.

Council of the European Union, General Secretariat. 1994. 1746th Council Meeting—General Affairs—Luxembourg, 18 and 19 April 1994. Press Release 6294/94, Brussels.

Cresson, E. 1985. "French Attitude to a New GATT Round." *The World Economy* 8(3): 317–19.

Curran, B., P. Minnis, and J. Bakalor. 1987. "Australian Agriculture in the International Community." *Quarterly Review of the Rural Economy* 9(1): 89–100.

Curzon, G. 1965. *Multilateral Commercial Diplomacy*. London: Michael Joseph.

Curzon, G., and V. Curzon, 1976. "The Management of Trade Relations in the GATT." In *International Economic Relations of the Western World, 1959–*

1971. Vol. 1, *Politics and Trade*. Edited by A. Shonfield. London: Oxford University Press for the Royal Institute of International Affairs.

Dam, K. W. 1970. *The GATT: Law and International Economic Organization*. Chicago: University of Chicago Press.

European Community. 1990. Uruguay Round. Agriculture. European Communities Offer Submitted Pursuant to MTN.TNC/15, unpublished (but widely circulated) document dated 7 November 1990.

European Council. 1992. *European Council in Edinburgh, 11–12 December 1992*. Conclusions of the Presidency. Luxembourg: Office for Official Publications of the European Communities.

Eurostat. 1973. *Yearbook of Agricultural Statistics 1973*. Luxembourg: Office for Official Publications of the European Communities.

Evans, J. W. 1971. *The Kennedy Round in American Trade Policy*. Cambridge: Harvard University Press.

Evans, P., and J. Walsh. 1994. *The EIU Guide to the New GATT*. London: Economist Intelligence Unit.

Financial Times. Various issues.

Food and Agriculture Organization (FAO). 1973. *Agricultural Protection: Domestic Policy and International Trade*. C73/LIM/9. Rome: Food and Agriculture Organization.

———. Various issues. *Food Outlook*. Rome: Food and Agriculture Organization.

French Government. 1993. Blair House, unpublished memorandum, 26 August 1993 (see also detailed summaries in *Agra Europe,* 3 and 17 September 1993).

GATT Focus. Various issues. Geneva: Information and Media Relations Division of GATT.

General Agreement on Tariffs and Trade (GATT). 1979. *The Tokyo Round of Multilateral Trade Negotiations*. Geneva: General Agreement on Tariffs and Trade.

———. 1989. Negotiating Group on Agriculture. *Global Proposals of the European Community on the Long-Term Objectives for the Multilateral Negotiations on Agricultural Questions*. MTN.GNG/NG5/W/145, 20 December 1989. Geneva: GATT Secretariat.

———. 1991. Trade Negotiations Committee, *Draft Final Act Embodying the Results of the Uruguay Round of Multilateral Trade Negotiations*. MTN.TNC/W/FA, 20 December 1991. Geneva: GATT Secretariat.

———. 1993. Negotiating Group on Market Access, *Modalities for the Establishment of Specific Binding Commitments under the Reform Programme, Note by the Chairman of the Market Access Group*. MTN.GNG/MA/W/24, 20 December 1993. Geneva: GATT Secretariat.

———. 1994a. *Agreement on Agriculture*. MTN/FA II-A1A-3. Geneva: GATT Secretariat.

———. 1994b. *Agreement on the Application of Sanitary and Phytosanitary Measures*. MTN/FA II-A1A-4. Geneva: GATT Secretariat.

Golt, S. 1978. *The GATT Negotiations 1973–79: The Closing Stage*. London: British–North American Committee.

Greenaway, D. 1991. "The Uruguay Round of Multilateral Trade Negotiations: Last Chance for GATT?" *Journal of Agricultural Economics* 42(3): 365–79.

Gummer, J. 1991. Text of speech to the National Farmers' Union. AGM, Ministry of Agriculture, Fisheries and Food news release, 46/91, 12 February 1991. London: Ministry of Agriculture, Fisheries and Food.

Harris, S. A. 1994a. "The Food Industry Perspective." In *Agriculture in the Uruguay Round*. Edited by K. A. Ingersent, A. J. Rayner, and R. C. Hine. London: Macmillan.

———. 1994b. "The Uruguay Round Outcome: A Dilemma for EU Food and Agriculture Policy." Paper prepared for the Thirty-Sixth Seminar of the European Association of Agricultural Economists, "Food Policies and the Food Chain: Structures and Inter-Relationships," 19–21 September 1994, The University of Reading, Reading, UK.

Harris, S. A., and A. Swinbank. 1991. "Dried Grapes: A Case Study in EC Market Disruption." *British Food Journal* 93(9): 10–16.

Harris, S. A., A. Swinbank, and G. A. Wilkinson. 1983. *The Food and Farm Policies of the European Community*. Chichester, UK: John Wiley.

Hartwig, B., T. E. Josling, and S. Tangermann. 1989. *Design of New Rules for Agriculture in the GATT*. Prepared, under a cooperative research agreement with NCFAP/Resources for the Future, for USDA and USTR. University of Göttingen, Göttingen, Germany.

Harvey, D. R. 1994. "Agricultural Policy Reform after the Uruguay Round." In *Agriculture in the Uruguay Round*. Edited by K. A. Ingersent, A. J. Rayner, and R. C. Hine. London: Macmillan.

Hathaway, D. E. 1987. *Agriculture and the GATT: Rewriting the Rules*. Policy Analyses in International Economics No. 20. Washington, D.C.: Institute for International Economics.

Herlihy, M. T., J. W. Glauber, and J. G. Vertrees. 1993. "U.S.-EC Blair House Agreement." In *International Agriculture and Trade Reports Europe*. Washington, D.C.: United States Department of Agriculture.

Hillman, J. S. 1994. "The US Perspective." In *Agriculture in the Uruguay Round*. Edited by K. A. Ingersent, A. J. Rayner, and R. C. Hine. London: Macmillan.

Hine, R. C., K. A. Ingersent, and A. J. Rayner. 1989. "Agriculture in the Uruguay Round: From the Punta del Este Declaration to the Geneva Accord." *Journal of Agricultural Economics* 40(3): 385–96.

Home-Grown Cereals Authority. 1994. "The GATT Agriculture Agreement." *Marketing Note. Supplement to Weekly Bulletin* 29(5).

Hudec, R. E. 1975. *The GATT Legal System and World Trade Diplomacy*. New York: Praeger.

Industry Commission. 1994. *Meat Processing*. Vol. 1, Report No. 38. Melbourne: Australian Government Publishing Service.

Ingersent, K. A., A. J. Rayner, and R. C. Hine. 1994. "The EC Perspective." In *Agriculture in the Uruguay Round*. Edited by K. A. Ingersent, A. J. Rayner, and R. C. Hine. London: Macmillan.

International Wheat Council. 1991. *World Grain Statistics*. London: International Wheat Council.

Johnson, D. G. 1973. *World Agriculture in Disarray*. London: Macmillan.

————. 1975. "World Agriculture, Commodity Policy, and Price Variability." *American Journal of Agricultural Economics* 57(5): 823–28.

————. 1987. "World Agriculture in Disarray Revisited." *Australian Journal of Agricultural Economics* 31(2): 142–53.

————. 1991. *World Agriculture in Disarray* 2d ed. London: Macmillan.

Josling, T. E. 1990. "The GATT: Its Historical Role and Importance to Agricultural Policy and Trade." In *The Political Economy of Agricultural Trade and Policy*. Edited by H. J. Michelmann, J. C. Stabler, and G. G. Storey. Boulder, CO: Westview.

————. 1991. "The CAP and the United States." In *The Common Agricultural Policy and the World Economy*. Edited by C. Ritson and D. R. Harvey. Wallingford, UK: CAB International.

————. 1993. "The Reformed CAP and the Industrial World." VIIth European Association of Agricultural Economists Congress, Stresa, Italy, 6–10 September 1993, Plenary Papers. Stresa: European Association of Agricultural Economists.

Kingma, O. 1987. "Performance of the Farm Sector." Paper presented to the National Agricultural Outlook Conference held by the Bureau of Agricultural Economics, 28–30 January 1987, Canberra, Australia.

Koester, U., et al. 1987. *Disharmonies in US and EC Agricultural Policy Measures*. Brussels: Commission of the European Communities.

Manegold, D. 1989. "EC Agricultural Policy in 1988–89: An Early End to Reform?" *Review of Marketing and Agricultural Economics* 57(1): 11–46.

McGovern, E. 1986. *International Trade Regulation: GATT, the United States, and the European Community*. 2d ed. Exeter, UK: Globefield Press.

Miller, G. 1987. *The Political Economy of International Agricultural Policy Reform*. Canberra: Australian Government Publishing Service.

Montagnon, P. 1990. "US Fast-Track Could Derail Deal." *Financial Times*, 4 December 1990, p. 3.

Organisation for Economic Co-operation and Development (OECD). 1987. *National Policies and International Trade*. Paris: Organisation for Economic Co-operation and Development.

————. 1988. *Monitoring and Outlook 1988*. Paris: Organisation for Economic Co-operation and Development.

Oxley, A. 1990. *The Challenge of Free Trade*. London: Harvester Wheatsheaf.

Paarlberg, R. 1991. *Why Agriculture Blocked the Uruguay Round: Evolving Strategies of a Two Level Game*. Cambridge: Harvard Center for International Affairs.

Palmeter, D. 1993. "Environment and Trade: Much Ado About Little?" *Journal of World Trade* 27(3): 55–70.

Peters, G. H. 1988. "The Interpretation and Use of Producer Subsidy Equivalents." *Oxford Agrarian Studies* 17: 186–218.

Petrey, L. A., and R. W. M. Johnson. 1993. "Agriculture in the Uruguay Round: Sanitary and Phytosanitary Measures." *Review of Marketing and Agricultural Economics* 61(3): 433–42.

Pryke, J., and D. Woodward. 1994. *The GATT Agreement on Agriculture: Will It Help Developing Countries?* Catholic Institute for International Relations Seminar Background Paper. London: Catholic Institute for International Relations.

Riethmuller, P., I. Roberts, L. P. O'Mara, G. Tie, V. Tulpule, M. Hossain, and N. Klijn. 1990. *Proposed Strategies for Reducing Agricultural Protection in the GATT Uruguay Round.* Discussion Paper 90.6. Australian Bureau of Agricultural and Resource Economics. Canberra: Australian Government Publishing Service.

Ritson, C., and A. Swinbank. 1986. *EEC Fruit & Vegetables Policy in an International Context.* Agra Europe Special Report No. 32. Tunbridge Wells, UK: Agra Europe (London) Ltd.

Roberts, I., G. Love, H. Field, and N. Klijn. 1989. *U.S. Grain Policies and the World Market.* Policy Monograph No. 4. Canberra: Australian Bureau of Agricultural and Resource Economics.

Roessler, F. 1987. "The Scope, Limits and Functions of the GATT Legal System." In *Trade Policies for a Better Future: The 'Leutwiler Report', the GATT, and the Uruguay Round,* 71–85. Dordrecht: Martin Nijhoff.

Snyder, F. G. 1985. *Law of the Common Agricultural Policy.* London: Sweet & Maxwell.

Stoeckel, A. 1985. *Intersectoral Effects of the CAP: Growth, Trade, and Unemployment.* Bureau of Agriculture Economics Occasional Paper No. 95. Canberra: Australian Government Printing Service.

Sumner, D. A. 1992. "The Economic Underpinnings of Uruguay Round Proposals." In *Improving Agricultural Trade Performance under the GATT.* Edited by T. Becker, R. Gray, and A. Schmitz. Kiel: Wissenschaftsverlag Vauk Kiel KG. (Papers presented at the conference "Mechanisms to Improve Agricultural Trade Performance under the GATT," 28–29 October 1991, Kiel).

Swinbank, A. 1989. "The Common Agricultural Policy and the Politics of European Decision Making." *Journal of Common Market Studies* 27(4): 303–22.

———. 1992. "GATT, Mac Sharry, and CAP Reform: Green Boxes, Amber Boxes, or Just Empty Boxes?" Paper presented at the conference "The Mac Sharry Reforms and the Future of the CAP," organized by the University Association for Contemporary European Studies and the European Public Policy Institute of the University of Warwick, 26 March 1992.

———. 1993a. "CAP Reform, 1992." *Journal of Common Market Studies* 31(3): 359–72.

———. 1993b. *The EC's New Agri-Monetary System.* Discussion Paper No. 93/04. Department of Agricultural Economics and Management, The University of Reading, Reading, UK.

———. 1994. "GATT, CAP, and the Manager's Dilemma." In *Farm Business*

Data 1993. Department of Agricultural Economics and Management, The University of Reading, Reading, UK.

Swinbank, A., and C. Ritson. 1995. "The Impact of the GATT Agreement on EU Fruit and Vegetable Policy." *Food Policy* 20(4): 339–57.

Swinbank, A., and C. Tanner. 1993. "Agriculture, the European Community, and the Uruguay Round." Paper presented at the thirty-seventh annual conference of the Australian Agricultural Economics Society at the University of Sydney, 9–11 February 1993.

Tangermann, S. 1991. "A Bond Scheme for Supporting Farm Incomes." In *The Changing Role of the Common Agricultural Policy: The Future of Farming in Europe.* J. Marsh, B. Green, B. Kearney, L. Mahé, S. Tangermann, and S. Tarditi. London: Belhaven.

Tangermann, S., and T. E. Josling. 1994. "The GATT and Community Preference for Cereals." *Agra Europe*, 15 July, E8–E9.

Tangermann, S., T. E. Josling, and S. Pearson. 1987. "Multilateral Negotiations on Farm-Support Levels." *The World Economy* 10(3): 265–81.

Tanner, C., and A. Swinbank. 1987. "Prospects for Reform of the Common Agricultural Policy." *Food Policy* 12(4): 290–94.

Trade Policies for a Better Future: The 'Leutwiler Report', the GATT, and the Uruguay Round 1987. Dordrecht: Martin Nijhoff.

Tweeten, L. 1986. "A Note on Explaining Farmland Price Changes in the Seventies and Eighties." *Agricultural Economics Research* 38(4): 25–30.

Tyers, R., and K. Anderson. 1992. *Disarray in World Food Markets. A Quantitative Assessment.* Cambridge: Cambridge University Press.

United States Department of Agriculture (USDA), Economic Research Service. 1987. *Government Intervention in Agriculture: Measurement, Evaluation, and Implications for Trade Negotiations.* FAER-229. Washington, D.C.: United States Department of Agriculture.

———. 1988a. *Estimates of Producer and Consumer Subsidy Equivalent: Government Intervention in Agriculture, 1982–86.* Staff Report AGES880127. Washington, D.C.: United States Department of Agriculture.

———. 1988b. *Agriculture in the Uruguay Round: Analysis of Government Support.* Staff Report AGES880802. Washington, D.C.: United States Department of Agriculture.

Viatte, G. 1990. "Agricultural Policies in OECD Countries: Agenda for the Future." *Journal of Agricultural Economics* 41(3): 292–302.

Warley, T. K. 1976. "Western Trade in Agricultural Products." In *International Economic Relations of the Western World, 1959–1971.* Vol. 1, *Politics and Trade.* Edited by A. Shonfield. London: Oxford University Press for the Royal Institute of International Affairs.

———. 1994. "The Canadian Perspective." In *Agriculture in the Uruguay Round.* Edited by K. A. Ingersent, A. J. Rayner, and R. C. Hine. London: Macmillan.

Winham, G. R. 1986. *International Trade and the Tokyo Round Negotiations.* Princeton, NJ: Princeton University Press.

172   References

Winters, L. A. 1987. "Reciprocity." In *The Uruguay Round: A Handbook on the Multilateral Trade Negotiations*. Edited by J. M. Finger and A. Olechowski. Washington, D.C.: The World Bank.
———. 1990. "The Road to Uruguay." *Economic Journal* 100: 1288–303.
World Bank. 1986. *World Bank Development Report 1986*. London: Oxford University Press.

Index